Maths
Foundation
Teacher's Guide

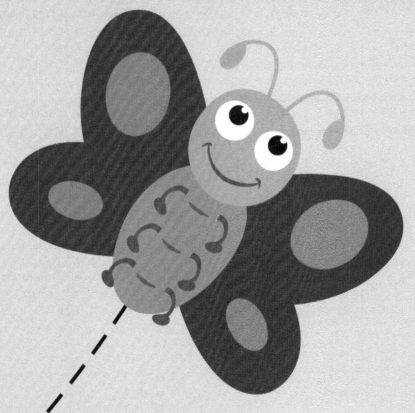

Peter Clarke

William Collins' dream of knowledge for all began with the publication of his first book in 1819.
A self-educated mill worker, he not only enriched millions of lives, but also founded a flourishing publishing house.
Today, staying true to this spirit, Collins books are packed with inspiration, innovation and practical expertise.
They place you at the centre of a world of possibility and give you exactly what you need to explore it.

Collins. Freedom to teach.

Published by Collins
An imprint of HarperCollins*Publishers*
The News Building, 1 London Bridge Street, London, SE1 9GF, UK

HarperCollins Publishers
Macken House, 39/40 Mayor Street Upper, Dublin 1, D01 C9W8, Ireland

> Browse the complete Collins catalogue at
> **www.collins.co.uk**

British Library Cataloguing-in-Publication Data
A catalogue record for this publication is available from the British Library.

Author: Peter Clarke
Publisher: Elaine Higgleton
Product manager: Letitia Luff
Commissioning editor: Rachel Houghton
Edited by: Sally Hillyer
Editorial management: Oriel Square
Cover designer: Kevin Robbins
Cover illustrations: Jouve India Pvt. Ltd.
Internal illustrations: p 56 Tasneem Amiruddin,
p 193 Adrita Das, p194 Lisa Williams; Jouve India Pvt. Ltd.
Typesetter: Jouve India Pvt. Ltd.
Production controller: Lyndsey Rogers
Printed and bound in the UK by Ashford Colour Press Ltd.

MIX
Paper | Supporting
responsible forestry
FSC™ C007454

This book is produced from independently certified FSC™ paper to ensure responsible forest management.

For more information visit: www.harpercollins.co.uk/green

Acknowledgements

With thanks to all the kindergarten staff and their schools around the world who have helped with the development of this course, by sharing insights and commenting on and testing sample materials:

Calcutta International School: Sharmila Majumdar, Mrs Pratima Nayar, Preeti Roychoudhury, Tinku Yadav, Lakshmi Khanna, Mousumi Guha, Radhika Dhanuka, Archana Tiwari, Urmita Das; Gateway College (Sri Lanka): Kousala Benedict; Hawar International School: Kareen Barakat, Shahla Mohammed, Jennah Hussain; Manthan International School: Shalini Reddy; Monterey Pre-Primary: Adina Oram; Prometheus School: Aneesha Sahni, Deepa Nanda; Pragyanam School: Monika Sachdev; Rosary Sisters High School: Samar Sabat, Sireen Freij, Hiba Mousa; Solitaire Global School: Devi Nimmagadda; United Charter Schools (UCS): Tabassum Murtaza; Vietnam Australia International School: Holly Simpson

The publishers wish to thank the following for permission to reproduce photographs.

(t = top, c = centre, b = bottom, r = right, l = left)

p 5t Rawpixel.com/Shutterstock, p 5b KAY4YK/Shutterstock, p 6tl Beloborod/Shutterstock, p 6tr ziggy_mars/Shutterstock, p 6bl DGLimages/Shutterstock, p 6br Boisvert/Shutterstock, p 7 DGLimages/Shutterstock, p 21 Marius Pirvu/Shutterstock, p 182 hobbit/Shutterstock, p 190–1 Shutterstock, p 192 thappypixelvectorstudio/Shutterstock, p 192b Brailescu Cristian/Shutterstock, p 196 Steve Lumb, p 220 Serhiy Kobyakov/Shutterstock

The publishers wish to thank the following for permission to reproduce text:

A Counting We Will Go', lyrics by Brian Haner, copyright © 2007 Music, Movement & Magination, Inc.; lyrics amended by Peter Clarke and used with the permission of Music, Movement & Magination, Inc. All rights reserved.

> Audio recordings available at:
> **www.collins.co.uk/internationalresources**

Contents

Introduction

Teaching notes

An introduction to kindergarten

About kindergarten children

Foundation children are creative, expressive and can often seem to be in their own world. Their emotions can be expressed by impulsive gestures and actions, as words may be lacking. If they are overwhelmed by emotion, they may need time out, or a quiet place to sit. They often play alongside each other (parallel play) although they are starting to understand the feelings of others.

Foundation children can't really attend to more than one thing at a time, or decide which situation needs attention the most (self-regulation). They may have a preferred schema of play (for example, repeating actions, moving things from place to place, covering things up, putting things into containers, moving in circles, throwing things.) However, they might not have the motor skills and coordination to sit still for long periods of time. They learn by mimicking your gestures and movements to develop their own. They love pattern and repeating actions. They enjoy music, dance, rhythm and rhymes, and often make their own rhythmic movements.

What are their needs?

A predictable and reassuring environment

A clear rhythm and structure to the day

Several hours a day for physical play, especially outdoors

Help with dressing and using the toilet

Food and rest at the right time

The kindergarten teacher

Foundation teachers observe, can sit back, and are sensitive to children's responses. They understand that behaviour isn't 'bad'. They are responsive to situations which may lead to new and unexpected learning that isn't in the lesson plan. They give positive feedback and modelling, and are people the children can rely on.

The kindergarten environment

When you enter a kindergarten classroom, there is usually a space for children to hang up their coats on their own hook, and a locker or shelf in which they can store their bags and outdoor shoes if used.

The classroom has desks, arranged in small groups, for drawing, painting and writing work. Each child should have their own seat in the group. This is the place to which they will return when asked to get ready for the next activity, for quiet time or to be counted. The teacher usually has a signal, like clapping three times, which lets the children know it is time to return to their seats. In some schools, there is a messy-play section with mud, clay, water or paint, and a place to wash up. There may also be a separate area for resting time.

The classroom has a reading corner – a carpeted area, sometimes with cushions, where the class or a group can gather for reading aloud or shared reading. This may be where the class's small library is kept.

Children also go to the reading corner to look at books in their own time. They can also go there to relax and for time out if they need a break from being with other children.

Outdoor play is a vital part of children's development. In the early years, children are highly active and are sometimes best able to demonstrate their achievements outdoors. They should spend at least three hours a day outside (not all of them at school), running, jumping, climbing, crawling, building, balancing and stretching. They should be digging holes, collecting stones and leaves, splashing in puddles, building forts and making dams.

Gross motor activities help to build muscle and coordination. There is a direct link between the development of coordination and motor skills and outdoor play. After all, the word *kindergarten* means 'a garden for children'. It was first used by Friedrich Froebel in 1840 to describe the school he envisaged, where children would learn through nature and the importance of play.

Activity stations

Most kindergarten classrooms have different activity stations, where children can play with sand, water, blocks and toy bricks, use playdough and clay, do construction activities, and enjoy small-world or imaginative play.

Introduction

Here are some activity stations with suggestions of useful materials and equipment for them.

Activity station	Material and equipment	Additional notes
Sand station (sandpit or tray)	Sticks, combs and rakes for mark-making. Buckets, spoons, spades, containers of various sizes, sieves, moulds, shells, flags and other sand play toys.	Cover the sandpit or tray after the children leave school in order to protect it from use by animals. It should also be cleaned regularly.
Waterplay station	Commercial waterplay trolley or large bowls or trays of water placed on crates or low tables. Different-sized containers, watering cans, buckets, plastic bottles, sieves and colanders, corks, straws, sponges, plastic aprons.	If possible, do waterplay outdoors where it doesn't matter if the children spill or splash water. If your waterplay station is inside, you will also need towels and a non-slip mat. Change the water every day. Do not use glass containers. Never leave children unsupervised when they are playing with water.
Construction station	Large carpeted area or mat, storage baskets or containers, building blocks, interlocking cubes, recycled boxes and packaging, toy construction bricks.	The area needs to be large enough for a small group of children to work alongside each other.
Playdough station	Playdough, base boards, playdough equipment (for example, rolling pins, shape cutters, modelling tools).	There are many easy playdough recipes on the internet based on flour, sugar and salt. You can alternate playdough with clay.
Painting and drawing station	Different-sized sheets of paper, crayons, coloured pencils, felt-tip pens, paint, brushes, aprons.	Large sheets of paper are best for painting.
Imaginative play station (Home corner)	Puppets, soft toys, dolls, plastic teacups and plates, blankets, fabric, dressing-up clothes and shoes, bags, toy cars, garage, doctor's set.	Create different environments linked to the unit's theme for children to count, sort, compare and explore.
Counting collections station	Small-world resources, e.g. people, animals and sea life, transport, double-sided counters, bottle tops, pebbles, pine cones, shells, matchsticks, lolly sticks, beads, buttons, corks, cotton reels, pegs, sorting trays, sorting rings, bowls, cups and pots, bags and containers for storage.	It is important to include an extensive counting collection, with different types of items sorted separately. Similarly, it is important to have a range of different sorting trays, hoops and other similar apparatus.

In some schools these may be permanent stations. In smaller schools, the teacher may need to set up the stations needed for each unit. (See the 'You will need' list at the start of each unit in the teaching notes.)

Other resources

Other resources that are useful in a kindergarten environment are:

- a display table for each theme, with objects the children can touch and investigate, for example, fruit and vegetables for the theme 'Food', or photos of each child for the theme 'Me'
- a weather chart on which the daily weather is recorded, displayed at child eye level (children can assist with this for the theme 'Weather')
- the day and date displayed at child eye level
- an alphabet display, with a different letter focused on each week, as well as a collection of objects that start with that letter
- number bunting or a large number track showing the numerals 1 to 10 and a picture to match each number (e.g. 4 dots below the number 4)
- a noticeboard outside the classroom where you can post notices and information for parents and guardians about what their children may need to bring to class.

What is *Collins Foundation?*

Aims and approach

Collins English, Maths and Science Foundation and Foundation Plus provide a complete course for children's learning in kindergarten. The course supports children in all the prior learning they will need to move successfully into their first year of primary education and is based on extensive research into how international schools structure teaching and learning in their kindergarten classes. The *Collins Foundation* programme offers progression through 15 integrated units, each suitable for two weeks' work and using popular themes. The sessions follow a pattern so that both you and the children will quickly become familiar with routines and expectations. The course follows an experiential, play-based approach and is aligned with the Early Years Foundation Stage areas of learning (England curriculum). It supports the requirements of key international curricula, including Cambridge Assessment International Education.

Components of the *Maths Foundation* course

Anthology

A full-colour stimulus book with fun images for teaching concepts in each of the units.

Introduction

Activity Books

Full-colour activity pages practising maths skills and embedding knowledge. Each book includes simple teaching notes and its own tracking page for the skills covered as a child's record of achievement.

Digital Tools

Tool	Functionality
Counting	Different scenes and objects to assist with counting from 1 (or 0) to 5, 10 or 20 and in teaching addition and subtraction.
Number track	Adaptable number track to assist with counting on and back and in teaching addition and subtraction.
Number cards	Different sets of number cards (numerals, dots and ten-frames) to assist with counting and in teaching addition and subtraction.
Tree	Objects to assist with counting and in teaching addition and subtraction.
Shape set	2D and 3D shapes to assist with recognising, comparing, sorting and creating repeating patterns and sequences.
Beads and laces	Beads and laces to assist with creating repeating patterns and sequences.

Flexible Digital Tools designed for use on an interactive whiteboard (IWB).

Slides

*Slides are provided for
many of the sessions as
visual aids.*

Teacher's Guide

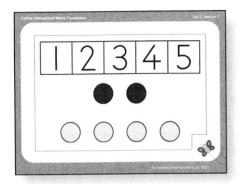

*A comprehensive and easy-to-use Teacher's Guide. An introduction to
kindergarten and thorough teaching notes with everything you need for
successful maths sessions using all the resources. Guidance and proforma
grids for assessment, plus photocopy masters, number fluency games and
activities, rhymes and songs and generic games.*

Photocopy masters (PCMs)

When using cards from a PCM
when teaching to the whole
class or large groups, it is best
to print the PCM onto A3 paper.

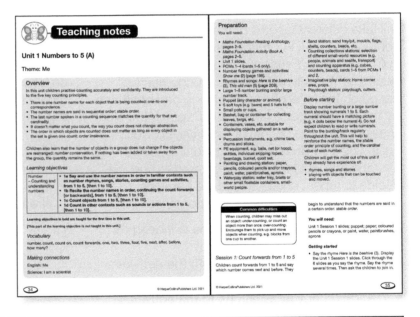

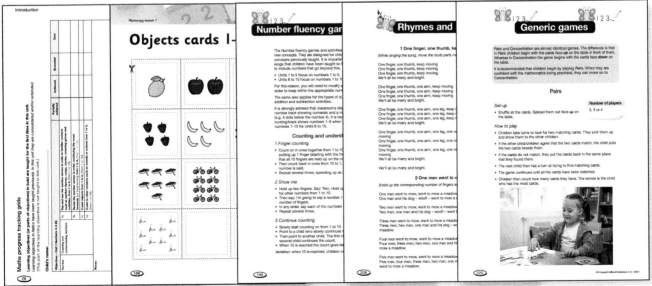

Using *Collins Maths Foundation*

Planning teaching and learning

The activities in *Collins Maths Foundation* take into account that young children need lots of opportunities to:

- explore, observe and discover things for themselves
- investigate and solve problems that are relevant to them
- practise and apply what they have learned.

You should dedicate time each day for purposeful mathematics activity. This should focus on supporting children to develop specific mathematical ideas and skills. Individual classes will take different lengths of time to learn the content of a unit. For this reason, each unit in *Collins Maths Foundation* consists of eight sessions. You will need to make judgements, based on the level of understanding demonstrated by the children, as to how you will spread the eight sessions over the course of ten days. For some units, you may decide to teach eight sessions over the two weeks; while for other units you may decide to repeat one or two of the sessions, or leave out some of the sessions altogether.

As you prepare for each unit, you will need to:

- make sure you are familiar with the activities in each part of the session
- decide on any adaptations to the activities that you need to make for your classroom setting
- gather, prepare and occasionally make the learning resources needed
- learn the songs, rhymes and games you will use well enough so that you do not need to refer to the written words or instructions during the session.

Grouping the children

For the 'Explore in groups' activities you will need to divide the children into groups.

At this age children should not be grouped by ability. Social grouping is far more beneficial and important in ensuring active learning for all children. It is a good idea to change the groups once or twice a term so that the children get the benefit of working with and learning from different peers.

Managing a maths session

Each session is divided into three parts: 'Getting started', 'Teaching' and 'Explore in groups'. 'Getting started' and 'Teaching' combined should last approximately 15 to 20 minutes, and 'Explore in groups' a similar length of time. However, this will vary depending on the needs and abilities of your class, the content of the session, and when in the academic year the session is being taught.

As children become more familiar with the structure of a session, they will increasingly be able to self-manage the transitions between the different parts.

About 'Getting started'

These activities are teacher-led. You will usually work with the whole class. However, depending on how your class is structured, you may wish to repeat the activity several times throughout the day to different groups of children. These activities:

- get the session off to a clear start and focus children's attention on maths
- may be used to quickly check what children already know about the topic you are about to teach
- may begin by:
  - discussing pages from the *Reading Anthology*
  - singing a song or saying a rhyme
  - playing a Number fluency game or activity.

About 'Teaching'

The 'Teaching' activities are teacher-led. You will usually work with the whole class. However, depending on how your class is structured, you may wish to repeat the activity several times throughout the day to different groups of children. However, regardless of how your class is organised, these activities:

- teach children a specific mathematical concept or skill, or consolidate previously taught concepts or skills
- include appropriate questions to take learning forward and check understanding
- should involve all children actively participating.

About 'Explore in groups'

For this part of the session the children are in groups at the activity stations, and work either independently or with adult support. The activities:

- encourage children to take their learning further and apply it in a new or different way
- have often been designed for the children to do independently of an adult
- should be led by the children, and any adult input should focus on observing, supporting and extending children's thinking
- often do not require collaborative group work even though the children may be seated in groups
- encourage conversation and discussion between the children as they work
- include *Activity Book* pages for some sessions, which should be completed with adult support.

Some have '(adult-led)' after them. These activities may benefit from an adult working closely with the children. This may be due to the complexity of the activity, or because intervention will help to develop conceptual understanding and mathematical language.

Some things to consider:

- The activities could be set up just for use during the maths session, or could remain for part or all of a day for children to access.
- An adult could work with the children one day on an activity, then put out the same resources for 'free play' on other days.
- There should be no expectation that all children will complete all of the activities in a particular session. Although children should be encouraged to explore the activities for themselves, adults should also direct individual children to the activities they feel are best suited to their needs and abilities. There is no expectation that all, or indeed any, of the 'Explore in groups' activities provided are used in a session. Use your professional judgement to decide how many activities to set up, and which activities will best meet your children's needs and abilities. You should also draw on your own repertoire of tried-and-tested activities.

Always supervise the children carefully. Be extra vigilant if they are using small parts.

Note

It is not necessary for children to read the words in either the *Maths Foundation Reading Anthology* or the *Activity Books*. Use these as read-aloud resources and read the words to the children, sometimes inviting them to try and predict what the words say when you have given sufficient clues to help them make reasonable guesses.

The role of the teaching assistant

Teaching assistants (or a second teacher) can support you and the children in the following ways:

- Helping to keep all the children focused during whole-class 'Getting started' and 'Teaching' activities.
- Helping children to complete *Activity Book* pages during 'Explore in groups'.
- Assisting children who are working at the activity stations so that they do not disturb you while you are working with other children.
- Gently helping to refocus children at the activity stations if needed. This can be done by finding out from the child what their intention is, then relating this back to the session topic. For example:
  Teaching assistant: *That looks interesting. What are you painting?*
  Child: *I'm painting a picture of a doll.*
  Teaching assistant: *That's interesting. Remember we were looking at things that are above or below other things. Maybe you could also paint something that is above or below the doll. What can you see that you might like to draw?*
- Leading (or repeating) the 'Getting started' and/or 'Teaching' activities in one or more sessions so that you can focus on observing the children for assessment purposes.

Introduction

Summary

Overview

In this unit children practise counting accurately and confidently. They are introduced to the five key counting principles.

- There is one number name for each object that is being counted: *one-to-one correspondence*.
- The number names are said in sequential order: *stable order*.
- The last number spoken in a counting sequence matches the quantity for that set: *cardinality*.
- It doesn't matter what you count, the way you count does not change: *abstraction*.
- The order in which objects are counted does not matter as long as every object in the set is given one count: *order irrelevance*.

Summary of different learning opportunities in the unit

Learning objectives

Number – Counting and understanding numbers	• **1a Say and use the number names in order in familiar contexts such as number rhymes, songs, stories, counting games and activities, from 1 to 5,** [then 1 to 10]. • **1b Recite the number names in order, continuing the count forwards** [or backwards], **from 1 to 5,** [then 1 to 10]. • **1c Count objects from 1 to 5,** [then 1 to 10]. • **1d Count in other contexts such as sounds or actions from 1 to 5,** [then 1 to 10].

Learning objectives in bold are taught for the first time in this unit.

[This part of the learning objective is not taught in this unit.]

Unit-specific learning objectives that feed into the learning objectives for the Maths Foundation course

Words introduced and/or used in the unit

Vocabulary

number, count, count on, count forwards, one, two, three, four, five, next, after, before, how many?

Making connections

English: Me

Science: I am a scientist

Links with English and science themes and/or concepts and skills

Preparation

You will need:

- *Maths Foundation Reading Anthology*, pages 2–9.
- *Maths Foundation Activity Book A*, pages 2–5.
- Unit 1 slides.
- PCMs 1–4 (cards 1–5 only).
- Number fluency games and activities: *Show me* (2) (page 198).
- Rhymes and songs: *Here is the beehive* (3), *This old man* (5) (page 209).
- Large 1–5 number bunting and/or large number track.
- Puppet (any character or animal).
- 5 soft toys (e.g. bears) and 5 hats to fit.
- Small pots or cups.
- Basket, bag or container for collecting leaves, twigs, etc.
- Containers, vases, etc. suitable for

- Sand station: sand tray/pit, moulds, flags, shells, counters, beads, etc.
- Counting collections stations: selection of different small-world resources (e.g. people, animals and sealife, transport) and counting apparatus (e.g. cubes, counters, beads), cards 1–5 from PCMs 1 and 2.
- Imaginative play station: Home corner area, props.
- Playdough station: playdough, cutters.

Before starting

Display number bunting or a large number track showing numerals 1 to 5. Each numeral should have a matching picture (e.g. 4 dots below the numeral 4). Do not

A list of resources that you will need to gather and/or prepare

Session number and title along with a brief description of the purpose of the session

> **Session 1: Count forwards from 1 to 5**
>
> Children count forwards from 1 to 5 and say which number comes next and before. They begin to understand that the numbers are said in a certain order: *stable order*.

A list of resources needed for the session

> **You will need:**
>
> Unit 1 Session 1 slides; puppet; paper; coloured pencils or crayons, or paint, water, paintbrushes, aprons

Specific preparation needed prior to the session and/or a whole-class activity ideas for introducing the session theme

> **Getting started**
>
> • Say the rhyme *Here is the beehive* (3). Display the Unit 1 Session 1 slides. Click through the 6 slides as you say the rhyme. Say the rhyme several times. Then ask the children to join in.

> **Teaching**
>
> • Say the numbers 1 to 5 in order. Then ask the children to count on from 1 to 5 with you. Repeat several times. Say: *We have counted on, or forwards, from 1 to 5.* Emphasise the words 'on' and 'forwards'.

A whole-class teaching activity, broken down into clear steps with suggested statements and questions to support you in achieving the purpose of the session

> **Explore in groups**
>
> **Music makers**
>
> • Spread out cards 1–5 from PCM 3 (cubes) face down. Encourage children to play a game in pairs or groups. Children take turns to turn over a card. They count the cubes and make that number of sounds using a percussion instrument. The other children count the sounds, checking that the number matches the number of cubes on the card.

Taking the learning further by exploring and practising individually, in pairs or in groups

> **Assessment opportunities**
>
> Assess children's learning against the objectives for this unit, using the guidance on formative assessment on pages 24–25, and record your observations in the Unit 1 progress tracking grid on page 26. The relevant pages of *Activity Book A* can also be used for assessment.

What you should look for in terms of achievement or understanding in this unit

Introduction

Recommended teaching and learning sequence

The table below shows the recommended teaching and learning sequence (often referred to as a medium-term plan) for the 15 units in *Collins* *Maths Foundation*. It is based on an academic year of three terms/semesters, each 10 weeks long. However, the units can be taught in any order to best meet the requirements and needs of individual schools, teachers and children.

	Term 1	**Term 2**	**Term 3**
Weeks 1 and 2	Unit 1 Numbers to 5 (A)	Unit 6 Numbers to 10 (A)	Unit 11 Numbers to 10 (B)
Weeks 3 and 4	Unit 2 Numbers to 5 (B)	Unit 7 Addition as combining two sets	Unit 12 Subtraction as taking away
Weeks 5 and 6	Unit 3 Position, direction and movement	Unit 8 Addition as counting on	Unit 13 Subtraction as counting back
Weeks 7 and 8	Unit 4 Length and height	Unit 9 Patterns and data	Unit 14 Mass and capacity
Weeks 9 and 10	Unit 5 2D shapes	Unit 10 Time	Unit 15 3D shapes

Developing number fluency beyond a maths session

It is important to find other opportunities throughout the day to create meaningful ways for children to use, practise and apply maths. This includes highlighting maths at various times of the day such as registration time, circle time, transitions, snack times, mealtimes and tidying up, as well as during play, and in other curriculum areas. For example, you could engage children in mathematical discussions, sing rhymes or songs, discuss pages from the *Reading Anthology*, or play a game or activity to develop number fluency.

Collins Maths Foundation includes a bank of Number fluency games and activities (pages 198–207). They help to reinforce key mental maths skills such as counting, comparing and ordering numbers, and addition and subtraction. These games and activities are adult-led and designed to be used with the whole class or a large group of children. They require no, or very few, resources, and are all oral activities rather than written. Each game or activity is designed to last no more than 10 minutes.

Many of the activities include variations which provide additional variety and choice. They also allow for adaptations to make activities easier or more challenging, or to broaden the activity to include other related concepts and skills.

Use the Number fluency games and activities often to continually revise and rehearse number concepts and skills, and to make sure the activities stay 'alive' in children's minds. The more familiar the activities become, the more confident children will feel. Although they may not realise it, they will be gaining fluency, honing their skills and improving their understanding of maths.

The activities could be used:

- as part of 'Getting started' to introduce a maths session
- at the end of the maths session to bring the session to a close
- at the start of the school day or just before home time
- before or immediately after play
- with quick-finishers while waiting for other children
- while waiting to go to assembly or similar
- at any free moment during the day
- with children's families, having sent the instructions and any resources home.

Teaching young children maths

Collins Maths Foundation is written to reflect the importance of:

- starting from what children know already and helping them to move their thinking forwards
- children learning the fundamental principles of an idea and making connections between different ideas
- focusing on the process of learning, rather than the end product of it
- developing children's thinking, by asking them questions and encouraging them to ask questions of their own
- learning through discovery and problem solving
- using collaborative, as well as individual, activities, so children can learn from each other

- evaluating the level of a child's development so suitable tasks can be set.

Concrete-pictorial-abstract (CPA) approach

Children can find maths difficult because it is often abstract. *Collins Maths Foundation* follows the CPA approach which builds on children's existing knowledge by introducing new concepts using physical objects or practical resources and equipment.

The CPA approach involves moving from **concrete** materials, to **pictorial** representations, to **abstract** symbols and problems. Deep understanding of a concept is achieved by going back and forth between these representations and making children aware of the connections between each one.

Concrete representation

The child is first introduced to a concept using physical objects. This hands-on approach is the foundation for conceptual understanding.

Pictorial representation

The child has understood the hands-on experiences and can relate them to images, such as a picture, diagram or model.

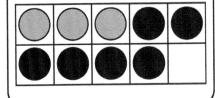

Abstract representation

The child is now capable of using numbers, notation and mathematical symbols to represent the concept.

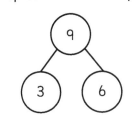

Thinking, reasoning and working mathematically

Mathematical development does not only involve conceptual understanding, knowledge and skills across a range of maths themes. It also involves being able to think, reason and work mathematically. This means encouraging children to:

- actively engage with their learning
- talk with others, and be able to explain their ideas
- develop reasoning skills such as logical thinking
- try to make sense of ideas

- build connections between different facts, skills and concepts
- gain meaning and satisfaction from their maths learning
- view the world in a mathematical way.

We do not want children to simply follow instructions and carry out processes that they have learned. We want them to know *why* the processes work and what the results *mean*. We want them to begin to see how maths applies to the real world and also to develop the skills needed to become independent mathematicians.

Thinking, reasoning and working mathematically should be embedded throughout sessions in every unit. Use the prompting questions

Introduction

below to draw out the characteristics that encourage young children to think, reason and work mathematically. The key feature of each characteristic is highlighted in **bold**.

Characteristic	Definition	Prompting questions
Classifying	**Comparing** and **contrasting** related ideas, **sorting** objects into groups according to their mathematical properties, **and explaining choices**.	*What's the same …?* *What's different …?* *What do these … have in common?* *How could you sort … into groups?* *How have you sorted/organised/grouped …?*
Reasoning	**Using logical thinking** and actions, recognising, explaining and justifying conclusions using appropriate language.	*What can you tell me about …?* *Why do you think that?* *Why does that happen?* *What if …?*
Generalising	**Recognising and explaining** underlying **patterns** by identifying examples.	*What do you notice?* *Can you show me an example of …?* *What is another example of …?* *What do you think might happen …? Why?*
Convincing	Explaining and **presenting evidence to justify or prove** a mathematical idea or solution.	*Does that work? Why/Why not?* *Is that right? Why/Why not?* *How do you know …?* *Tell/Show me why …*
Communicating	Organising and **explaining thinking** using mathematical vocabulary, concrete resources, pictorial representations, personal jottings and symbolic notation.	*Show me …* *Tell me …* *Why?* *Explain to me …* *How do you know …?*
Problem-solving	**Using and applying mathematics** to understand, represent, solve and interpret and evaluate problems.	*Does this work? Why/Why not?* *What's the problem?* *Is this right? Why/Why not?* *What do you know that can help you?*
Connecting	**Identifying relationships** within and between different maths themes, between maths learning and learning in other subjects, and between maths and its application to the real world.	*What do you notice?* *How is this the same as …?* *How is this different from …?* *How is this like …?* *What do you know that will help you?* *You know … so what else do you know?*

The teaching of money

Given the international nature of *Collins Maths Foundation*, and the different currencies used, there is no unit covering the topic of money.

You could occasionally use coins in your own currency instead of other counting apparatus (e.g. counters or cubes). Only use coins that have the number 1 written on them (e.g. 1c, $1, £1, €1), so that cardinal and monetary values match.

Key concepts in kindergarten maths

The following are the key concepts young children should explore in kindergarten. They are fundamental to lifelong mathematical thinking. They link numerous mathematical ideas into a coherent whole, and should guide all teaching and learning experiences in the early years.

Counting and understanding numbers

The idea of counting can appear simple. However, to be able to count accurately and confidently, children need to develop a set of complex principles.

- There is one number name for each object that is being counted: *one-to-one correspondence*.
- The number names are said in sequential order: *stable order*.
- The last number spoken in a counting sequence matches the quantity for that set: *cardinality*.
- It doesn't matter what you count, the way you count does not change: *abstraction*.
- The order in which objects are counted does not matter as long as every object in the set is given one count: *order irrelevance*.
- The number of objects in a group does not change if objects are rearranged: *number conservation*.
- Quantities and numbers can be compared and ordered by 'more than', 'less than' and 'equal to'.

Using fingers for maths

Children should be encouraged to use their fingers while they are developing their counting skills and early understanding of addition and subtraction.

Fingers can help make abstract maths concepts and procedures more concrete and easier to understand. Fingers also have the advantage of always being available! The total number of fingers is also equivalent to a very special number in maths: 10.

In addition, using fingers helps children to understand and practise:

- saying the number names in sequential order: *stable order* principle
- the cardinal value of numbers to 10 by pairing a number gesture with its corresponding number word, for example, assigning the word *three* to the gesture of holding up 3 fingers
- the counting-on addition strategy
- the counting-back subtraction strategy.

Patterns

Almost all mathematics is based on pattern – a predictable sequence. Seeking and exploring patterns should be at the heart of all maths teaching and learning.

- A pattern is a repeating (or growing) sequence that is determined by a rule.
- To identify the rule of a pattern, children must make predictions. This leads children towards observing and explaining generalisations.
- The same pattern structure, like a repeating AB pattern, can be found in many different forms such as shape, colour, size and type.
- Patterns can be made with objects such as coloured cubes (visual patterns), as well as with actions and sounds, and they exist in the real-world as well as in maths.

Introduction

Addition and subtraction

The four number operations are the fundamental mathematical tools we use in everyday life. Addition and subtraction are the first two operations that children learn. Children will learn that an operation performed on a set of numbers tells a mathematical 'story'.

- A quantity (the 'whole') can be separated (decomposed) into equal or unequal parts. Parts can be combined (composed) to create a whole.
- Two (or more) quantities can be combined into a single quantity to work out the total (or sum) – putting together by counting all.
- A given quantity can be increased by another quantity – becoming greater by counting on.
- A smaller quantity can be taken away from a larger quantity to work out how many are left (or how much is left).
- A given quantity can be decreased (or reduced) by another quantity – becoming smaller by counting back.

Shape

Shape awareness is one of the first mathematical discoveries that young children make. However, their discoveries need to be made explicit so that they are able to recognise the similarities and differences between different shapes.

- Children in kindergarten will describe, define and classify 2D (flat) shapes by their sides, and 3D (solid) shapes by their faces.
- Shapes can be regular or irregular (e.g. 'tall skinny' triangles, 'short fat' triangles, right-angled triangles). Shapes can also be presented in different sizes, in different orientations and from different perspectives.
- The flat faces of 3D (solid) shapes are 2D (flat) shapes.

Spatial reasoning

Mathematically, 'space' is about representing and locating objects. Children need to develop an understanding of the relationship between objects, and start to gain spatial awareness.

- Describing and representing spatial relationships can differ depending on the orientation and perspective.
- Spatial awareness requires an understanding of the associated vocabulary. The vocabulary is both numerous and interconnected, and includes subtle differences in language.
- To describe and represent the relative positions of objects in a picture or diagram, children need to be able to interpret a 2D representation.
- Spatial reasoning incorporates spatial memory – knowing where things are, and a sense of direction – finding your way.

Measurement

Measurement is an essential mathematical procedure that we use and apply to many different real-world contexts.

- Measurement involves different attributes such as length, height, width, mass, capacity and volume.
- Different attributes can be measured, even when measuring the same object.
- The attributes of two or more objects can be *directly* compared by matching one item against another.
- The attributes of two or more objects can be *indirectly* compared using an intermediary device, such as a piece of string, to assist with the comparison.
- The quantity of an object is unchanged, even if it is rearranged or if its appearance is altered: *conservation*.

Sorting

Sorting involves grouping together a collection of items in some meaningful way into different sets. Children need to identify the similarities and differences between different collections. This will allow them to determine whether or not an item belongs in a particular set. Sorting also encourages mathematical reasoning and leads children to making generalisations.

- Objects can be compared and sorted into sets according to various criteria, such as shape, colour, size and type.
- The same collection of objects can be sorted in different ways.
- Sorting objects into sets leads to children developing an understanding of how we define a collection of objects.

Developing maths language

Part of learning new maths concepts and knowledge involves learning new language. Helping children develop the language to think and talk about maths is just as important as making sure that they are exposed to new skills and knowledge. Young children develop their maths vocabulary when they engage in hands-on practical activities set up to help them investigate knowledge and concepts. As they learn new words and practise using them, they deepen their maths knowledge too.

For children who do not speak English at home, this also means learning about maths concepts in a language they have not fully mastered.

Ideas to support speakers of English as a second language

- Constantly reinforce connections:

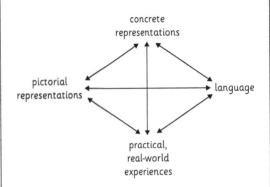

- Encourage non-verbal responses such as thumbs up or down, or using fingers to represent quantities/numbers.
- When asked to talk in front of others, allow children extra 'think time'.
- Encourage children to work in pairs or small groups with a confident English speaker.
- Promote partner talk.
- Constantly praise, encourage and provide positive reinforcement.
- When children work independently, allow extra time so that they complete a task at their own pace.
- When questioning the whole class, use a combination of targeted questions to specific children as well as asking for choral responses.

Learning objectives for *Collins Maths Foundation*

Below are the learning objectives for the *Collins Maths Foundation* course. Together with the *Foundation Plus* course, these will equip the children with the skills and understanding they will need for their first year of primary education. The table also indicates in which units each objective is taught and/or practised.

The learning objectives are organised into three main areas called 'strands': Number, Geometry and Measure, and Statistics. Within each strand there is a number of 'sub-strands'.

NUMBER		Units
Counting and understanding numbers		
1a	Say and use the number names in order in familiar contexts such as number rhymes, songs, stories, counting games and activities, from 1 to 5, then 1 to 10.	1, 2, 6, 11
1b	Recite the number names in order, continuing the count forwards or backwards, from 1 to 5, then 1 to 10.	1, 2, 6, 11
1c	Count objects from 1 to 5, then 1 to 10.	1, 2, 6, 11
1d	Count in other contexts such as sounds or actions from 1 to 5, then 1 to 10.	1, 2, 6, 11
1e	Share objects into two equal groups.	11
Reading and writing numbers		
2a	Recognise numerals from 1 to 5, then 1 to 10.	2, 6, 11
2b	Begin to record numbers, initially by making marks, progressing to writing numerals from 1 to 5, then 1 to 10.	2, 6, 11
Comparing and ordering numbers		
3a	Use language such as more, less or fewer to compare two numbers or quantities from 1 to 5, then 1 to 10.	2, 6, 11
Understanding addition and subtraction		
4a	In practical activities and discussions begin to use the vocabulary involved in addition: combining two sets and counting on.	7, 8
4b	In practical activities and discussions begin to use the vocabulary involved in subtraction: take away and counting back.	12, 13
4c	Find 1 more and 1 less than a number from 1 to 10.	8, 13
Patterns and sequences		
5a	Talk about, recognise and make simple patterns using concrete materials or pictorial representations.	5, 9

GEOMETRY AND MEASURE	Units
Understanding shape	
6a	Identify, describe, compare and sort 2D shapes.	5
6b	Identify, describe, compare and sort 3D shapes.	15
Position, direction and movement	
7a	Begin to understand and use the vocabulary of position, direction and movement.	3
Measurement	
8a	Use everyday language to describe and compare length, height and width, including long, longer, short, shorter, tall, taller, wide, wider, narrow and narrower.	4
8b	Use everyday language to describe and compare mass including heavy, heavier, light and lighter.	14
8c	Use everyday language to describe and compare capacity and volume including more, less, full and empty.	14
Time	
9a	Begin to understand and use the vocabulary of time, including the days of the week, yesterday, today, tomorrow, morning and evening.	10
9b	Sequence familiar events.	10

STATISTICS	Units
10a	Sort, represent and describe data using concrete materials or pictorial representations.	5, 9, 15

Assessment

While it is possible to identify broad child development milestones, individual children's learning and development are largely dependent on individual strengths and length of exposure to stimulating environments. The focus of assessment in the kindergarten classroom should be to create a map of how far individual children have come in what they can do, think about and express. This makes informal formative assessment (which focuses on what children have achieved and are able to do, rather than what they cannot do) the most appropriate assessment for kindergarten classrooms. Not only does this type of assessment describe children's performance over time, but it also helps you to make decisions about the best way to support children's ongoing development.

Tracking children's progress

Track children's progress by gathering evidence over time so that you don't need to rely on memory when it comes to reporting on their progress. It is best to use many different ways of gathering this evidence. For example:

- Observe children during the whole class, teacher-led 'Getting started' and 'Teaching' activities, and independent 'Explore in groups' activities. You can record your observations using the tracking grids (pages 26–33) as well as writing them in a notebook. (A notebook is especially useful for recording children's achievements that fall outside of the intended learning objectives of a unit.)
- Record children's understanding of specific maths concepts, knowledge and skills at the end of each unit. You can use the tracking grids (pages 26–33) as well as page 24 of each *Activity Book* for this.
- Listen carefully to the questions children ask you and the conversations they have with their peers. These provide a window on the depth of children's understanding of concepts and topics.
- Look carefully at the objects that children sort or create, their drawings, any writing/pretend-writing, and their completed *Activity Book* tasks. These show you what they understand and have achieved.

Feedback

Use the box at the bottom of each Activity Book page to either:

- sign or initial your name to indicate you have assessed the page
- draw a simple picture or diagram such as one of the three faces

- If you are granted permission officially by parents and guardians, use a camera or a camera app on a mobile phone to take photographs and/or videos of children as they engage in different activities. This is very useful to share with parents/guardians during teacher-parent consultations and helps to provide examples of the conclusions you reach about children's progress at a given point in the year.

Assessment tips

- Formal testing of kindergarten children is an inappropriate method of gathering evidence of their progress.
- Observation is best done while children are engaged in usual classroom activities.
- While you are busy observing a small group of children, the other children could be engaged in completing an *Activity Book* task or in activities at the activity stations.

Using the tracking grids

The purpose of the tracking grids on pages 26–33 is to:

- record each child's progress in meeting the learning objectives necessary to prepare them for their next year of education
- help generate useful reports on children's progress for parents/guardians and principals.

Photocopy the relevant page to allow one grid per pupil.

The following table shows how the descriptions of children's achievement on the tracking grids relate to the skills and understanding in the learning objectives.

	Learning objectives related to skills	Learning objectives related to understanding
Partially achieved	Used when a child is some way to mastering the skill but relies on some adult assistance to complete the task.	Used when a child shows evidence of partially understanding a concept or topic and needs time and further engagement to fully grasp it.
Achieved	Used when a child is able to perform the skill independently of an adult.	Used when a child shows evidence of having understood a concept or topic.
Exceeded	Used when the child performs the skill at a more advanced level than described in the learning objective.	Used when a child shows evidence of having a deeper or broader understanding of a concept or topic than is expressed in the learning objective.

The progress tracking grids list all the learning objectives taught in each unit.

Once a decision has been made regarding a child's level of skill and understanding in a particular learning objective, tick the appropriate column: Partially achieved, Achieved or Exceeded. Write the date on which you made each assessment.

Then in the 'Notes' section for each unit, write any specific comments that you feel are appropriate and give some indication of the overall performance level that the child has achieved in the unit as a whole.

Reporting on children's progress

All teachers are accountable to the principal of their school and the children's parents/guardians for what children learn in their classrooms and the progress they make towards preparing for their next year of education. The assessment evidence that you collect can be used for the completion of a formal written progress report each term or semester. It can also be used in parent-teacher consultations and should referrals for specialist interventions be necessary for some children.

Parents and guardians often welcome more frequent and informal feedback on their children's progress. This is important because it encourages their involvement in supporting their children's learning at school. Letting children take home their *Activity Book* at the end of each unit so that parents/guardians can see and ask their child about their completed tasks, as well as take note of your recorded observations on page 24, helps to provide them with feedback. In addition, creating an in-classroom display to showcase any objects the children have made and drawings/paintings they have done during a unit also provides an opportunity to strengthen parents'/guardians' interest in children's school life and builds good home-school relationships that nurture children.

Maths progress tracking grids

Learning objectives (or parts of objectives) in bold are taught for the first time in this unit.

Learning objectives in italics have been taught previously. In this unit they are consolidated and/or extended.

[This part of the learning objective is not taught in this unit.]

Child's name: _____

Objectives – Unit 1 Numbers to 5 (A)			Partially achieved	Achieved	Exceeded	Date
Number	Counting and understanding numbers	1a	**Say and use the number names in order in familiar contexts such as number rhymes, songs, stories, counting games and activities, from 1 to 5,** [then 1 to 10].			
		1b	**Recite the number names in order, continuing the count forwards** [or backwards], **from 1 to 5,** [then 1 to 10].			
		1c	**Count objects from 1 to 5,** [then 1 to 10].			
		1d	**Count in other contexts such as sounds or actions from 1 to 5,** [then 1 to 10].			
Notes:						

Child's name: _____

Objectives – Unit 2 Numbers to 5 (B)

			Partially achieved	Achieved	Exceeded	Date
Number	Counting and understanding numbers	1a	*Say and use the number names in order in familiar contexts such as number rhymes, songs, stories, counting games and activities, from 1 to 5, [then 1 to 10].*			
		1b	*Recite the number names in order, continuing the count forwards or backwards, from 1 to 5, [then 1 to 10].*			
		1c	*Count objects from 1 to 5, [then 1 to 10].*			
		1d	*Count in other contexts such as sounds or actions from 1 to 5, [then 1 to 10].*			
	Reading and writing numbers	2a	**Recognise numerals from 1 to 5, [then 1 to 10].**			
		2b	**Begin to record numbers, initially by making marks, progressing to writing numerals from 1 to 5, [then 1 to 10].**			
	Comparing and ordering numbers	3a	**Use language such as more, less or fewer to compare two numbers or quantities from 1 to 5, [then 1 to 10].**			

Notes:

Objectives – Unit 3 Position, direction and movement

			Partially achieved	Achieved	Exceeded	Date
Geometry and Measure	Position, direction and movement	7a	**Begin to understand and use the vocabulary of position, direction and movement.**			

Notes:

Introduction

Child's name: _____

Objectives – Unit 4 Length and height

			Partially achieved	Achieved	Exceeded	Date
Geometry and Measure	Measurement	8a	Use everyday language to describe and compare length, height and width, including long, longer, short, shorter, tall, taller, wide, wider, narrow and narrower.			

Notes:

Objectives – Unit 5 2D shapes

			Partially achieved	Achieved	Exceeded	Date
Geometry and Measure	Understanding shape	6a	Identify, describe, compare and sort 2D shapes.			
Number	Patterns and sequences	5a	Talk about, recognise and make simple patterns using concrete materials or pictorial representations.			
Statistics	-	10a	Sort, represent and describe data using concrete materials or pictorial representations.			

Notes:

Child's name: _____

Objectives – Unit 6 Numbers to 10 (A)			Partially achieved	Achieved	Exceeded	Date
Number	Counting and understanding numbers	1a	Say and use the number names in order in familiar contexts such as number rhymes, songs, stories, counting games and activities, from 1 to 5, **then 1 to 10.**			
		1b	Recite the number names in order, continuing the count forwards or backwards, from 1 to 5, **then 1 to 10.**			
		1c	Count objects from 1 to 5, **then 1 to 10.**			
		1d	Count in other contexts such as sounds or actions from 1 to 5, **then 1 to 10.**			
	Reading and writing numbers	2a	Recognise numerals from 1 to 5, **then 1 to 10.**			
		2b	Begin to record numbers, initially by making marks, progressing to writing numerals from 1 to 5, **then 1 to 10.**			
	Comparing and ordering numbers	3a	Use language such as more, less or fewer to compare two numbers or quantities from 1 to 5, **then 1 to 10.**			

Notes:

Objectives – Unit 7 Addition as combining two sets			Partially achieved	Achieved	Exceeded	Date
Number	Understanding addition and subtraction	4a	**In practical activities and discussions begin to use the vocabulary involved in addition: combining two sets** [and counting on].			

Notes:

Introduction

Child's name: _____

Objectives – Unit 8 Addition as counting on			Partially achieved	Achieved	Exceeded	Date
Number	Understanding addition and subtraction	4a	*In practical activities and discussions begin to use the vocabulary involved in addition: [combining two sets and] counting on.*			
		4c	**Find 1 more [and 1 less] than a number from 1 to 10.**			

Notes:

Objectives – Unit 9 Patterns and data			Partially achieved	Achieved	Exceeded	Date
Number	Patterns and sequences	5a	*Talk about, recognise and make simple patterns using concrete materials or pictorial representations.*			
Statistics	-	10a	*Sort, represent and describe data using concrete materials or pictorial representations.*			

Notes:

Child's name: _____

Objectives – Unit 10 Time

				Partially achieved	Achieved	Exceeded	Date
Geometry and Measure	Time	9a	Begin to understand and use the vocabulary of time, including the days of the week, yesterday, today, tomorrow, morning and evening.				
		9b	Sequence familiar events.				

Notes:

Objectives – Unit 11 Numbers to 10 (B)

				Partially achieved	Achieved	Exceeded	Date
Number	Counting and understanding numbers	1a	Say and use the number names in order in familiar contexts such as number rhymes, songs, stories, counting games and activities, from 1 to 5, then 1 to 10.				
		1b	Recite the number names in order, continuing the count forwards or backwards, from 1 to 5, then 1 to 10.				
		1c	Count objects from 1 to 5, then 1 to 10.				
		1d	Count in other contexts such as sounds or actions from 1 to 5, then 1 to 10.				
		1e	Share objects into two equal groups.				
	Reading and writing numbers	2a	Recognise numerals from 1 to 5, then 1 to 10.				
		2b	Begin to record numbers, initially by making marks, progressing to writing numerals from 1 to 5, then 1 to 10.				
	Comparing and ordering numbers	3a	Use language such as more, less or fewer to compare two numbers or quantities from 1 to 5, then 1 to 10.				

Notes:

Child's name: _____

Objectives – Unit 12 Subtraction as taking away

Number			Partially achieved	Achieved	Exceeded	Date
Understanding addition and subtraction	4b	In practical activities and discussions begin to use the vocabulary involved in subtraction: take away [and counting back].				

Notes:

Objectives – Unit 13 Subtraction as counting back

Number			Partially achieved	Achieved	Exceeded	Date
Understanding addition and subtraction	4b	In practical activities and discussions begin to use the vocabulary involved in subtraction: [take away and] counting back.				
	4c	Find [1 more and] 1 less than a number from 1 to 10.				

Notes:

Child's name: _____

Objectives – Unit 14 Mass and capacity

			Partially achieved	Achieved	Exceeded	Date
Geometry and Measure	Measurement	8b	**Use everyday language to describe and compare mass including heavy, heavier, light and lighter.**			
		8c	**Use everyday language to describe and compare capacity and volume including more, less, full and empty.**			

Notes:

Objectives – Unit 15 3D shapes

			Partially achieved	Achieved	Exceeded	Date
Geometry and Measure	Understanding shape	6b	**Identify, describe, compare and sort 3D shapes.**			
Statistics	-	10a	*Sort, represent and describe data using concrete materials or pictorial representations.*			

Notes:

 **Teaching notes**

Unit 1 Numbers to 5 (A)

Theme: Me

Overview

In this unit children practise counting accurately and confidently. They are introduced to the five key counting principles.

- There is one number name for each object that is being counted: *one-to-one correspondence.*
- The number names are said in sequential order: *stable order.*
- The last number spoken in a counting sequence matches the quantity for that set: *cardinality.*
- It doesn't matter what you count, the way you count does not change: *abstraction.*
- The order in which objects are counted does not matter as long as every object in the set is given one count: *order irrelevance.*

Children also learn that the number of objects in a group does not change if the objects are rearranged: *number conservation.* If nothing has been added or taken away from the group, the quantity remains the same.

Learning objectives

Number – Counting and understanding numbers	• **1a Say and use the number names in order in familiar contexts such as number rhymes, songs, stories, counting games and activities, from 1 to 5, [then 1 to 10].** • **1b Recite the number names in order, continuing the count forwards [or backwards], from 1 to 5, [then 1 to 10].** • **1c Count objects from 1 to 5, [then 1 to 10].** • **1d Count in other contexts such as sounds or actions from 1 to 5, [then 1 to 10].**

Learning objectives in bold are taught for the first time in this unit.

[This part of the learning objective is not taught in this unit.]

Vocabulary

number, count, count on, count forwards, one, two, three, four, five, next, after, before, how many?

Making connections

English: Me

Science: I am a scientist

Preparation

You will need:

- *Maths Foundation Reading Anthology*, pages 2–9.
- *Maths Foundation Activity Book A*, pages 2–5.
- Unit 1 slides.
- PCMs 1–4 (cards 1–5 only).
- Number fluency games and activities: *Show me* (2) (page 198).
- Rhymes and songs: *Here is the beehive* (3), *This old man* (5) (page 209).
- Large 1–5 number bunting and/or large number track.
- Puppet (any character or animal).
- 5 soft toys (e.g. bears) and 5 hats to fit.
- Small pots or cups.
- Basket, bag or container for collecting leaves, twigs, etc.
- Containers, vases, etc. suitable for displaying objects gathered on a nature walk.
- Percussion instruments, e.g. chime bars, drums and sticks.
- PE equipment, e.g. balls, net (or hoop), skittles, individual skipping ropes, beanbags, bucket, quoit set.
- Painting and drawing station: paper, pencils, coloured pencils and/or crayons, paint, water, paintbrushes, aprons.
- Waterplay station: water tray, boats or other small floatable containers, small-world people.
- Sand station: sand tray/pit, moulds, flags, shells, counters, beads, etc.
- Counting collections stations: selection of different small-world resources (e.g. people, animals and sealife, transport) and counting apparatus (e.g. cubes, counters, beads), cards 1–5 from PCMs 1 and 2.
- Imaginative play station: Home corner area, props.
- Playdough station: playdough, cutters.

Before starting

Display number bunting or a large number track showing numerals 1 to 5. Each numeral should have a matching picture (e.g. 4 dots below the numeral 4). Do not expect children to read or write numerals. Point to the bunting/track regularly throughout the unit. This will help to reinforce the number names, the *stable order* principle of counting, and the *cardinal* value of each number.

Children will get the most out of this unit if they already have experience of:

- rhymes, songs and stories
- playing with objects that can be touched and moved.

Common difficulties

When counting, children may miss out an object: *under-counting*, or count an object more than once: *over-counting*. Encourage them to pick up and move objects when counting, e.g. blocks from one cup to another.

Session 1: Count forwards from 1 to 5

Children count forwards from 1 to 5 and say which number comes next and before. They begin to understand that the numbers are said in a certain order: *stable order*.

You will need:

Unit 1 Session 1 slides; puppet; paper; coloured pencils or crayons, or paint, water, paintbrushes, aprons

Getting started

- Say the rhyme *Here is the beehive* (3). Display the Unit 1 Session 1 slides. Click through the 6 slides as you say the rhyme. Say the rhyme several times. Then ask the children to join in.

Teaching notes

Teaching

- Say the numbers 1 to 5 in order. Then ask the children to count on from 1 to 5 with you. Repeat several times. Say: *We have counted on, or forwards, from 1 to 5.* Emphasise the words 'on' and 'forwards'.
- Say: *I have a friend who needs help with number names and counting.* Introduce the puppet.
- Ask the children to help the puppet to count on from 1. Lead them in chanting the numbers 1 to 5. Repeat several times.
- Emphasise the words 'next' and 'after': *The number after 2 is 3.*
- Ask: *Who can tell the puppet the number that comes after 1?* Repeat for other numbers.
- Do the same for the word 'before': *The number that comes before 4 is 3.*
- All count together forwards from 1 to 5. Repeat several times. Ask the children to count as loudly as they can, or whisper the numbers, or sing them, etc.

Explore in groups

Repeat *Teaching* (adult-led)

- Repeat the *Teaching* activity with smaller groups of children. Encourage each child to join in with counting on from 1 to 5. Say: *The puppet can't remember what number comes before 2 [or 3, 4 or 5]. Can you help?* Repeat for the number that comes *after* 1 (or 2, 3 or 4).

Puppet mistakes (adult-led)

- Explain that the puppet is going to count forwards from 1 to 5. Make the puppet say the numbers 1 to 5, but make a deliberate mistake, e.g. *1, 2, 3, 5.* Ask children to spot the mistake. Repeat several times. Each time, leave out a number, or say the same number twice, or say the numbers out of order.

Painting and drawing station

- Invite children to draw or paint a picture based on the rhyme *Here is the beehive.* Ask questions focused on the number of objects, e.g. *How many bees have you drawn? Can you remember how many are in the rhyme? Would you like to draw all five bees?*

Session 2: Count 1 and 2 objects

Children count groups of 1 and 2 objects.

You will need:

Maths Foundation Reading Anthology; 2 soft toys (e.g. bears) and 2 hats to fit; cards 1 and 2 from PCMs 1–4; selection of different small-world resources and counting apparatus; boats or other small floatable containers; sand tray, moulds, flags, shells, counters, beads, etc; paper; coloured pencils or crayons, or paint, water, paintbrushes, aprons

Getting started

- Show pages 2 and 3 (At home) of the *Reading Anthology.* Discuss the picture with the children. Point to, and count, groups of 1 and 2. Repeat for pages 4 and 5 (In the garden).

Teaching

- Tell children that today they will find out about counting things (objects).
- Show 1 soft toy (e.g. a bear): *I can see 1 bear.* Put a hat on to the bear: *I can see 1 hat.* Show cards 1 and 2 from PCM 1 (fruit). Ask children to point to the card with 1 fruit.
- Then show 2 bears and 2 hats. Point to each bear: *I can see 1, 2 bears. I can see 1, 2 hats.* Show fruit cards 1 and 2 again. Ask children to point to the card with 2 fruits.
- Show all the 1 and 2 cards from PCMs 1–4. Ask children to point to the cards with 1 object. Say each time: *1 finger; 1 cube; 1 dot;* etc.
- Repeat for groups of 2. All chant as you point to the objects: *1, 2: 2 fingers; 1, 2: 2 cubes;* etc.
- Repeat, counting groups of 1 and 2 small-world resources and counting apparatus.
- Repeat, counting groups of 1 and 2 objects linked to the theme 'Me': *1, 2 hands; 1, 2 eyes;* etc.

Explore in groups

Waterplay station

- Provide boats or other small floatable containers. Invite children to put 1 or 2 small-world people into each boat/container.

Sand station

- In the sand tray, provide moulds and objects such as flags, shells, counters, beads, etc. Invite children to make sandcastles, using the moulds. Encourage them to decorate their sandcastles with 1 or 2 of each object.

Painting or drawing station

- Provide sheets of paper, each with a large oval drawn on. Invite children to create a face: 1 mouth, 1 nose, 2 eyes, 2 ears and hair. They could draw, or paint using their fingers or a brush.

Maths Foundation Reading Anthology
(adult-led)

- Together, look at pages 2 and 3 of the *Reading Anthology*. Repeat the *Getting started* activity. Repeat for pages 4 and 5.

Session 3: Count up to 3 objects

Children count groups of up to 3 objects. They practise giving just one number name to each object: *one-to-one correspondence*.

You will need:

Maths Foundation Reading Anthology; 3 soft toys (e.g. bears) and 3 hats to fit; cards 1–3 from PCMs 1–4; selection of different small-world resources and counting apparatus; small pots or cups; paper; coloured pencils or crayons, or paint, water, paintbrushes, aprons; Home corner area and props

Getting started

- Show pages 6 and 7 (Goldilocks and the three bears) of the *Reading Anthology*. Discuss the picture with the children. Point to, and count, groups of 1, 2 and 3. Encourage children to join in with the counting.

Teaching

- As in *Session 2*, use 3 bears and 3 hats to introduce children to the number name 'three', and how this is represented by 3 objects.
- Show cards 1–3 from PCM 1 (fruit). Ask children to point to the card that shows 3 fruit.
- Then show a selection of cards 1–3 from PCMs 1–4. Ask children to point to the cards with 3 fruit/fingers/cubes/dots. Each time, reinforce 'three': *1, 2, 3. That's 3 fingers. 1, 2, 3. That's 3 cubes.*
- Emphasise giving one number name to each object in turn: *one-to-one correspondence*. Make sure that children do not count twice: *over-count*, or miss any out: *under-count*.
- Show groups of up to 3 different small-world resources and counting apparatus. Ask children to count the objects in each group. For example, point to a group of 3 toy cars: *How many cars? That's right. 1, 2, 3. There are 3 cars.*

Explore in groups

Counting collections station

- Invite children to make groups of 3 small-world resources and counting apparatus. They could place each group in a different pot, e.g. a pot of 3 dinosaurs, a pot of 3 farm animals.

Painting and drawing station

- Allow each child to choose an object to draw or paint. Ask them to draw 3 of their object, e.g. 3 boats, 3 cats.

Imaginative play station

- Depending on what is set up in the Home corner area, encourage children to do things in threes, e.g. dress 3 dolls, prepare dinner for 3 people, put 3 cars into each garage.

Maths Foundation Reading Anthology
(adult-led)

- Together, look at pages 6 and 7 of the *Reading Anthology*. Repeat the *Getting started* activity.

Teaching notes

Session 4: Count up to 4 objects

Children count groups of up to 4 objects. They begin to understand that the last number spoken matches the quantity for that set: *cardinality*.

You will need:

4 soft toys (e.g. bears) and 4 hats to fit; cards 1–4 from PCMs 1–4; selection of different small-world resources and counting apparatus; sand tray, moulds; red counters (or similar); basket, bag or container for collecting leaves, twigs, etc; containers, vases, etc. suitable for displaying the objects gathered on the nature walk; paper; coloured pencils or crayons, or paint, water, paintbrushes, aprons; playdough, cutters

Getting started

- Sing the action song *This old man* (5) (verses 1 to 4 only).

Teaching

- As in *Sessions 2 and 3*, use 4 bears and 4 hats to introduce children to the number name 'four', and how this is represented by 4 objects.
- Show cards 1–4 from PCM 1 (fruit). Ask children to point to the card that shows 4 fruit.
- Then show a selection of cards 1–4 from PCMs 1–4. Ask children to point to the cards with 4 fruit/fingers/cubes/dots. Each time, reinforce 'four': *1, 2, 3, 4. That's 4 dots. 1, 2, 3, 4. That's 4 fingers.*
- Ask questions such as: *How many apples are there? How many oranges are on this card? How many dots can you see?* This will help children to understand that the last number in the count tells you how many objects there are: *cardinality*. For example, when you ask *How many oranges are there?* children do not need to say, *1, 2, 3*, the answer is simply the last number in the count: *3*.
- Show groups of up to 4 different small-world resources and counting apparatus. Ask children to count the objects in each group.

Explore in groups

Sand station

- Let children make sand 'pies' using moulds. Encourage them to decorate each pie with 4 'cherries' (red counters or similar).

Nature walk collection (adult-led)

- Take the children outdoors for a nature walk. Ask them to collect leaves, pine cones, feathers, acorns, twigs, pebbles and other natural objects. In the classroom, ask children to make groups of up to 4 objects, and arrange each type in a vase/container. For example, 1 feather, 2 twigs, 3 acorns, 4 leaves.

Painting and drawing station

- Invite children to draw or paint a group of 4 people, including themselves.

Playdough station

- Invite children to use playdough and cutters to make objects. Prompt them to create groups of 4, e.g. 4 sheep, 4 pineapples, 4 trees. If cutters are not available, children could make 4 cookies, 4 balls, 4 snakes.

Session 5: Count up to 5 objects

Children count groups of up to 5 objects. They understand that anything can be counted: *abstraction*. They also begin to understand that it doesn't matter in which order we count a group of objects: *order-irrelevance*.

You will need:

5 soft toys (e.g. bears) and 5 hats to fit; cards 1–5 from PCMs 1–4; selection of different small-world resources and counting apparatus; *Maths Foundation Reading Anthology*; *Maths Foundation Activity Book A*; pencils

Getting started

- Use the Number fluency activity *Show me* (2). Focus on groups of up to 5 objects (fingers).

Teaching

- As in *Sessions 2 to 4*, use 5 bears and 5 hats to introduce children to the number name 'five', and how this is represented by 5 objects.
- Show cards 1–5 from PCM 2 (fingers). Ask the children to point to the card that shows 5 fingers.
- Then show a selection of cards 1–5 from PCMs 1–4. Ask the children to point to the

cards with 5 fruit/fingers/cubes/dots. Each time, reinforce 'five': *1, 2, 3, 4, 5. That's 5 cubes.*

- Explain that it doesn't matter what you count: *abstraction.* Say: *We can count fingers, fruit, cubes, dots, children, chairs – almost anything. We can even count sounds, things that move and things we can't see and touch.*
- Emphasise that the order in which objects are counted doesn't matter: *order-irrelevance.* For example, demonstrate counting a group of 5 toy animals or children, in different ways: left to right, right to left, top to bottom, bottom to top. Explain: *If you count every object just once, and don't miss out any, you will always get the same number.*
- Show groups of up to 5 different small-world resources and counting apparatus. Ask children to count the objects in each group.

Explore in groups

Pairs

- Provide sets of cards 1–5 from PCM 2 (fingers). Encourage children to play Pairs (see Generic games rules on page 220). Each pair or group will need two sets of cards.

Counting collections station

- Place cards 1–5 from PCM 2 next to a variety of small-world resources. Prompt children to make collections of resources to match the groups of fingers on the cards. Encourage children to check each other's groups by counting.

Maths Foundation Reading Anthology (adult-led)

- Together, look at pages 8 and 9 (The stream) of the *Reading Anthology*. Discuss the picture with the children. Point to and count groups of up to 5. Encourage the children to find and count other groups of up to 5 objects.

Maths Foundation Activity Book A (adult-led)

Page 2 – Count and match

Session 6: Count up to 5 children and objects

Children count groups of up to 5. They begin to understand that the number of objects in a group does not change if the objects are rearranged: *conservation.*

You will need:

cards 1–5 from PCMs 1 and 2; selection of different small-world resources and counting apparatus; *Maths Foundation Reading Anthology*; *Maths Foundation Activity Book A*; coloured pencils or crayons

Getting started

- Use the Number fluency activity *Show me* (2). Focus on groups of up to 5 objects (fingers).

Teaching

- Remind children that they have been learning number names and counting groups of 1 to 5 objects. *Point to 1 ear. Show me 2 hands. Show me 3/4/5 fingers.*
- Choose a child to stand up. Ask all the children: *How many children are standing?* (*1*) Choose a second child to stand next to the first. Ask the same question. Choose a third child to stand in a row with the first two. Ask the question again.
- Swap the positions of the children. Ask: *How many children are standing?*
- Now position the children so that they are not in a row, e.g. each child stands in a different area of the room. Again, ask: *How many children are standing?*
- Ask two of the children to sit down. Ask: *Now how many children are standing?* (*1*)
- Repeat, for 4 children, and then 5 children.
- Show cards 1–5 from PCMs 1 and 2. Ask children to count the objects in each group. Repeat for groups of up to 5 different small-world resources and counting apparatus.

Explore in groups

Pairs

- Provide sets of cards 1–5 from PCMs 1 (fruit) and 2 (fingers). Encourage children to play Pairs (see Generic games rules on page 220).

Teaching notes

Each pair or group will need one set of fruit cards and one set of finger cards.

Counting collections station

- Place cards 1–5 from PCM 1 (fruit) next to a variety of small-world resources. Prompt children to make collections of resources to match the groups of fruit on the cards. Encourage children to check each other's groups by counting.

Maths Foundation Reading Anthology (adult-led)

- Together, look at pages 2 and 3 (At home) of the *Reading Anthology*. Briefly discuss the picture, pointing to and counting groups of up to 5. Encourage the children to find and count other groups of up to 5 objects. Repeat for pages 4 and 5 (In the garden) and pages 6 and 7 (Goldilocks and the three bears).

Maths Foundation Activity Book A (adult-led)

Page 3 – Count and draw

Session 7: Count up to 5 sounds and actions

Children begin to understand that they can count non-physical things such as sounds, actions and remembered or imaginary objects: *abstraction*.

You will need:

percussion instruments, e.g. chime bars, drums and sticks; cards 1–5 from PCM 3; PE equipment; *Maths Foundation Activity Book A*; pencils

Getting started

- Sing the action song *This old man* (5) (verses 1 to 5 only). Discuss the numbers in the rhyme. Explain that numbers can be used for counting sounds and actions, not just objects.

Teaching

- Ask children to count (up to 5) things that they can see, e.g. doors, windows, tables, adults and so on.
- Explain how we can also count things that we cannot see. Clap your hands 3 times. Ask children to count the claps they hear.
- Repeat, varying the number (up to 5) and the type of sounds, e.g. knocking or playing a percussion instrument. Ask children to count first with their eyes open, then with their eyes closed.
- Then explain how we can also count actions (things that we do). Jump 3 times. Ask children to count the jumps.
- Repeat, varying the number (up to 5) and the type of action, e.g. taking big steps, hopping or tapping your head.
- Ask children to make a given number of sounds or actions, e.g. *Clap your hands 2 times; knock on the floor 5 times; tap your shoulders 3 times; stamp your feet 4 times.*

Explore in groups

Music makers

- Spread out cards 1–5 from PCM 3 (cubes) face down. Encourage children to play a game in pairs or groups. Children take turns to turn over a card. They count the cubes and make that number of sounds using a percussion instrument. The other children count the sounds, checking that the number matches the number of cubes on the card.

Outdoor play (adult-led)

- Create a simple obstacle course. Each 'obstacle' must be completed a given number of times, e.g. kick 1 ball into a net (or hoop); knock down 2 skittles; skip 3 times using a skipping rope; throw or drop 4 beanbags into a bucket; throw or drop 5 quoits over the peg.

Maths Foundation Activity Book A (adult-led)

Page 4 – Listen and count

Slowly clap your hands up to 5 times. Children count each clap and circle the matching number of pictures. Repeat for the other three actions.

Session 8: Count forwards to 5, not starting at 1

Children count forwards to 5 from a number other than 1. They say which number comes next and before.

You will need:

Unit 1 Session 1 slides; puppet; paper; coloured pencils or crayons; cards 2–5 from PCMs 1–4; *Maths Foundation Activity Book A*; coloured pencils or crayons

Getting started

- Say the rhyme *Here is the beehive* (3), to practise counting forwards from 1 to 5. Display the Unit 1 Session 1 slides, clicking through the six slides as you say the rhyme together.

Teaching

- Say the numbers 1 to 5 in order. Then ask the children to count on from 1 to 5 with you. Repeat several times. Say: *We have counted on, or forwards, from 1 to 5.* Emphasise the words 'on' and 'forwards'.
- Show the puppet. Ask children to help the puppet count from 1 to 5.
- Remind children of the words 'next' and 'after', e.g. *When we count on, the next number after 2 is 3.*
- Ask: *Who can tell the puppet the number that comes after 2?* Repeat for other numbers.
- Do the same for the word 'before': *The number that comes before 3 is 2.*
- Once children are confident counting forwards from 1 to 5, start the count at a number other than 1. *Let's count on to 5 again, but this time starting from 2. Ready? Let's go!*
- Repeat several times. Sometimes ask all the children to count, and sometimes just individual children.

Explore in groups

Painting and drawing station

- Invite children to draw around their hand, then decorate the 'handprint' in any way they like. Ask them to tell you how many fingers they have drawn.

Sorting and ordering cards

- Choose a selection of cards 2–5 from PCMs 1–4. Shuffle them. Spread them out face up on the table. Children could sort the cards into four groups: fruit, fingers, cubes and dots. Then encourage them to put each set of cards in order, e.g. 3 cubes, 4 cubes, 5 cubes. Ask children to say the numbers in order for each set, e.g. 3, 4, 5.

Maths Foundation Activity Book A (adult-led)

Page 5 – Count and draw

Assessment opportunities

Assess children's learning against the objectives for this unit, using the guidance on formative assessment on pages 24–25, and record your observations in the Unit 1 progress tracking grid on page 26. The relevant pages of *Activity Book A* can also be used for assessment.

Can the children:

- show some evidence that they understand the five key counting principles?
- recite the number names 1 to 5 in order counting forwards, starting from 1 and then from a number other than 1?
- say which number comes next and which number comes before when counting forwards?
- count up to 5 objects accurately?
- recognise that the number of objects in a group does not change if the objects are rearranged: *conservation*?
- count up to 5 things that cannot be touched?

Unit 2 Numbers to 5 (B)

Theme: My family

Overview

This unit consolidates the five key counting principles introduced in Unit 1. Children need to understand these in order to count with accuracy and confidence.

These principles are extended in this unit. Children recite the number names 1 to 5 backwards. They start learning to read and write the numerals 1 to 5.

Children also learn the comparative language: 'more', 'less' (for uncountable nouns, measured quantities and numbers: *Sam has less flour than Di; 3 is less than 5*) and 'fewer' (for quantities that can be counted: *There are fewer oranges than bananas*).

Learning objectives

Number – Counting and understanding numbers	• *1a Say and use the number names in order in familiar contexts such as number rhymes, songs, stories, counting games and activities, from 1 to 5,* [then 1 to 10]. • *1b Recite the number names in order, continuing the count forwards* **or** **backwards***, from 1 to 5,* [then 1 to 10]. • *1c Count objects from 1 to 5,* [then 1 to 10]. • *1d Count in other contexts such as sounds or actions from 1 to 5,* [then 1 to 10].
– Reading and writing numbers	• **2a Recognise numerals from 1 to 5,** [then 1 to 10]. • **2b Begin to record numbers, initially by making marks, progressing to writing numerals from 1 to 5,** [then 1 to 10].
– Comparing and ordering numbers	• **3a Use language such as more, less or fewer to compare two** **numbers or quantities from 1 to 5,** [then 1 to 10].

Learning objectives in italics have been taught previously. In this unit they are consolidated and/or extended.

Learning objectives (or parts of objectives) in bold are taught for the first time in this unit.

[This part of the learning objective is not taught in this unit.]

Vocabulary

number, numeral, count, count on, count forwards, count back, count backwards, one, two, three, four, five, how many, more, less, fewer, the same, more than, less than

Making connections

English: My family

Science: Living and non-living

Preparation

You will need:

- *Maths Foundation Reading Anthology*, pages 8 and 9.
- *Maths Foundation Activity Book A*, pages 6–11.
- Unit 1 Session 1 slides.
- Unit 2 slides.
- Digital Tools: Tree, Number track, Counting.
- PCMs 1–5 (cards 1–5 only) and 6–8.
- Rhymes and songs: *Here is the beehive* (3), *Zoom, zoom, we're going to the moon* (8) (pages 209–211).
- Large 1–5 number bunting and/or large number track.
- Puppet (any character or animal).
- Stapler.
- Glue.
- Scissors.
- Percussion instruments, e.g. chime bars, drums and sticks.
- Small pieces of paper.
- Various classroom objects and displays that have numbers, e.g. a clock, number bunting, room number, wall displays.
- Magazines, newspapers, catalogues, etc.
- Poster paper.

- 5 hoops (or similar).
- At least 15 beanbags.
- Painting and drawing station: paper, pencils, coloured pencils and/or crayons, paint, water, paintbrushes, aprons.
- Counting collections station: a selection of different small-world resources (e.g. people, animals and sealife, transport) and counting apparatus (e.g. cubes, counters), small pots or cups, numeral cards 1–5 from PCM 5.
- Sand station: sand tray/pit, sets of plastic and/or wooden numerals 1–5.

Before starting

As for Unit 1, display number bunting or a large number track showing numerals 1 to 5. Each numeral should have a matching picture (e.g. 4 dots below the numeral 4). Refer to the bunting/track regularly throughout the unit.

Before teaching this unit, look back at the Assessment section at the end of Unit 1 (page 41), to identify the prerequisite learning for this unit.

Common difficulties

Children may take a long time to learn to write numerals correctly. At this stage, do not worry about errors (e.g. children writing numerals backwards). It is more important that children understand the *cardinal* value of each number. They may choose to represent a number in their own way. For example, a child might draw three small marks to represent the number 3.

Maths background

The word 'fewer' is used when talking about people or things in the plural. The word 'less' is used when talking about things that are uncountable, or have no plural. It is important to use the correct mathematical vocabulary throughout this unit. When comparing two sets of objects, use the words 'more' and 'fewer'. When comparing two numbers, use 'more' and 'less'.

Teaching notes

Session 1: Count backwards from 5

Children count backwards from 5 to 1.

You will need:

Unit 1 Session 1 slides; puppet; cards 1–5 from PCMs 1–4; paper; coloured pencils or crayons, or paint, water, paintbrushes, aprons

Getting started

- Say the rhyme *Here is the beehive* (3), to practise counting forwards from 1 to 5. Display the Unit 1 Session 1 slides. Click through the 6 slides as you all say the rhyme together.

Teaching

- Ask the children to count forwards from 1 to 5. Repeat several times.
- Slowly count backwards from 5 to 1. Repeat. Ask: *What do you notice about the way I counted this time? What was the same? What was different?*
- Discuss: *I used the same numbers to count. But instead of counting forwards from 1 to 5, I counted backwards from 5 to 1.* Emphasise the words 'forwards' and 'backwards'. Count backwards from 5 to 1 again.
- Show the puppet. Ask children to help the puppet to count backwards from 5 to 1. Repeat several times.
- Say the action rhyme *Zoom, zoom, we're going to the moon!* (8). Repeat several times. Then ask the children to join in.

Explore in groups

Puppet mistakes (adult-led)

- Explain that the puppet is going to count backwards from 5 to 1. Make the puppet say the numbers 5 to 1, but make a deliberate mistake, e.g. *5, 4, 2, 1*. Ask children to spot the mistake. Repeat several times. Each time, leave out a number, or say the same number twice, or say the numbers out of order.

Ordering cards

- Provide sets of cards 1–5 from PCM 2 (fingers). Shuffle the cards in each set. Encourage children to count the fingers on each card in their set. Ask: *Can you put them in a line, in order from 5 to 1?*

Variation: use cards 1–5 from PCMs 1, 3 or 4.

Painting and drawing station

- Invite children to draw or paint a picture based on the rhyme *Zoom, zoom, we're going to the moon* (8). Ask children to count down the rocket blast-off from 5 to 1.

Session 2: Compare two quantities to 5

Children compare two quantities. They are introduced to the words 'more', 'fewer' and 'the same'.

You will need:

Maths Foundation Reading Anthology; Tree Digital Tool; PCM 6; coloured pencils or crayons; cards 1–5 from PCM 1; 6 counters (or similar); selection of different small-world resources and counting apparatus; at least 20 small pots or cups; stapler or glue

Getting started

- Show pages 8 and 9 (The stream) of the *Reading Anthology*. Discuss the picture with the children. Point to different animals/objects (duck, snail, water strider, dragonfly, fish, water lily, flower, hat). Ask children to tell you how many there are.

Teaching

- Display the Tree Digital Tool. Set it to show 2 trees. Place 3 apples on the first tree and 2 apples on the second tree. Point and say: *There are 3 apples on this tree, and 2 apples on this tree. There are more apples on this tree. There are fewer apples on this tree.* Emphasise the words 'more' and 'fewer'.
- Click 'Clear all'. Repeat, placing up to 5 apples on each tree.
- Repeat, placing 3 apples on each tree. Point and say: *There are 3 apples on this tree, and 3 apples on this tree. There are the same number of apples on each tree.* Emphasise the phrase 'the same'.
- Repeat several times. Each time, place up to 5 apples on each tree (either a different amount on each tree, or the same). Ask children to count the apples on each tree. Ask them to say which tree has *more/fewer* apples, or

whether both trees have *the same* number of apples.

- Place 2 apples on the first tree. Say: *I want more than 2 apples on the second tree. How many apples could I put on this tree?* Agree that the answer could be 3, 4 or 5. Place 3, 4 or 5 apples on the second tree. Point and say: *There are 2 apples on this tree, and 3/4/5 on this tree. Which tree has more apples? Which tree has fewer apples?*
- Place up to 5 apples on the first tree. Ask children to suggest *more*, *fewer* or *the same* number of apples that could go on the second tree.

Explore in groups

More apples

- Give each child a copy of PCM 6. Tell each child a number from 1 to 4. Ask them to draw that many apples on the first tree. Then ask them to draw *more* apples on the second tree.

Variations: ask children to draw 2 to 5 apples on the first tree and then draw *fewer* apples (or *the same* number of apples) on the second tree.

Comparing fruit (adult-led)

- Shuffle two sets of cards 1–5 from PCM 1 (fruit). Spread the 10 cards face down on the table. Ask two children to each turn over a card. Ask: *Who has the card with more fruit?* Give the child with more fruit a counter (or similar). If two children turn over matching cards, and say that they are 'the same', give each child a counter. Continue until all 10 cards have been used.

Variation: ask: *Who has the card with fewer fruit?*

Counting collections station

- Provide 6 or 7 pots, each containing up to 5 small-world resources or counting apparatus. Also provide some empty pots and some loose resources/apparatus. Encourage children to take a full pot and count the number of resources inside. Then invite them to fill an empty pot with *more/fewer/the same* number of resources.

More or fewer cards

- Before the activity, prepare some double-sided cards. Use cards 1–5 from PCM 1 (fruit). Staple or glue pairs of cards back-to-back.

You will need at least 3 or 4 double-sided cards for each child. Invite children to count the fruit on each side of the card and circle the group that is *more*.

Variation: ask children to circle the group that is *fewer*.

Session 3: Use marks to record quantities 1 to 5

Children use mark-making to represent numbers.

You will need:

cards 1–5 from PCMs 1–4; selection of different small-world resources and counting apparatus; percussion instruments; 6 or 7 small pots or cups small pieces of paper; pencils; *Maths Foundation Reading Anthology*

Getting started

- Revise counting up to 5 objects. Use cards 1–5 from PCMs 1–4. Hold up a card. Ask children to tell you how many fruit/fingers/cubes/dots are on the card. Repeat several times.
- Occasionally hold up two cards. Ask the children to point to the card with more/fewer objects.

Teaching

- Say: *Sometimes it's hard to remember how many things we have counted. But there is a way to help us.*
- Show a set of up to 5 small-world resources or counting apparatus, e.g. 4 planes. As you point to each object, draw one large tally mark on the board as you say the number in the count. Repeat for another set of up to 5 objects.
- Repeat again. This time, draw small dots or rings on the board instead of tally marks. Discuss how you have used a different way to record the number of objects in the group. Explain that it doesn't matter how you show the amount, as long as it is simple, quick and easy to write.
- Remind children that we can also count things we cannot touch, such as sounds and movements. Ask a child to clap their hands 3 times. Make one mark on the board for each clap.

Teaching notes

- Repeat. Vary the type and number of sounds (up to 5). For example, use a percussion instrument.
- Repeat. Ask individual children to perform an action a certain number of times (up to 5), e.g. *Touch your nose 4 times.*

Explore in groups

Counting collections station

- Provide 6 or 7 pots, each containing up to 5 small-world resources or counting apparatus. Place some small pieces of paper next to the pots. Invite children to count the contents of each pot, making a mark on a piece of paper for each object they count. Prompt children to use a different piece of paper for each pot.

Counting and representing pictorial representations

- Give each child two or three different 1–5 cards from PCMs 1 and 2, and two or three small pieces of paper. Ask children to count the fruit/fingers on each card, making a mark on a piece of paper for each fruit/finger they count. Prompt children to use a different piece of paper for each card.

Maths Foundation Reading Anthology
(adult-led)

- Together, look at pages 8 and 9 (The stream) of the *Reading Anthology*. Give each child a small piece of paper. Tell each child (or point to) a different creature in the picture. Ask them to draw a small picture of their creature. Ask children to count how many of their creature they can see. For each one they count, they make a mark on their paper. If time allows, repeat. Ask children to compare their pieces of paper. Do all the pieces of paper with the same creature have the same number of marks? How are the marks the same/different?

Session 4: Recognise numerals 1 to 5
Children begin to recognise numerals up to 5.

You will need:

various classroom objects and displays that have numbers; Unit 2 Session 4 slide; magazines, newspapers and catalogues; poster paper; scissors; glue; sand tray; sets of plastic and/ or wooden numerals 1–5; selection of different small-world resources and counting apparatus; paper; coloured pencils or crayons

Getting started

- Tell children to watch carefully as you move around the classroom. As you move, point to some numbers (written as numerals) you come across.
- Return to the children. Ask: *What do you notice about what I did?* Discuss the fact that all of the objects have numbers.

Teaching

- Ask children to talk about their experiences of numbers: *What numbers do you know? Where have you seen numbers? What are they for? How are they used?*
- Explain: *Numbers are marks or signs used to show how many of something there is. They are all around us. We use numbers when we count and in many other ways.*
- Ask children to point to other numbers in the classroom and/or outdoors.
- Some children may be unsure of the number names. Encourage other children to name the numbers if they are able to do so.
- Discuss the use of numbers in the classroom, e.g. *The numbers on the clock help us tell the time. The numbers on our coat pegs help us to remember which peg is ours. The number track helps us to count, to see how a number is written, and also which number comes next and before.*
- Display slide 1. Point to each number in turn. Say the number name. Ask the children to repeat the number. Discuss each number, highlighting its 'shape'. If appropriate, discuss the context in which the number is used.

Explore in groups

Number display (adult-led)

- Provide poster paper, scissors, glue and a selection of magazines, newspapers and catalogues. Work with the children to create a poster/display of numbers. Focus on the numerals 1 to 5, but also include other numbers suggested by the children.

Sand station

- Before the activity, bury one or more sets of plastic and/or wooden numerals 1–5 in the sand tray. Also bury several different small-world resources and counting apparatus. Children aim to find one each of the numerals 1 to 5.

Number hunt (adult-led)

- Take the children on a hunt for numbers around the school (both indoors and outdoors). Each time they find a number, stop and discuss the number and its purpose. In the classroom, ask children to draw pictures of the numbers they saw.

Session 5: Match a numeral to a set of 1 to 5 objects

Children begin to use the numerals 1 to 5 to represent quantities.

You will need:

Number track Digital Tool; Counting Digital Tool; cards 1–5 from PCMs 1–5; selection of different small-world resources and counting apparatus; small pots or cups; 5 hoops (or similar); at least 15 beanbags; *Maths Foundation Activity Book A*; pencils

Getting started

- Display the Number track Digital Tool. Set it to show a 1 to 5 number track. Say: *These are the numbers 1 to 5.*
- Recite the numbers together, counting forwards from 1 to 5. Point to each number as you say it. Repeat several times.
- Then count backwards from 5 to 1. Again, point to each number as you say it. Repeat several times.

Teaching

- Display the Counting Digital Tool. Set it to show the beach scene with the crab and fish.
- Hold up numeral card 3 from PCM 5. Say: *I'm going to place some fish in the scene. When there are 3 fish, hold up 3 fingers and shout: '3!'*
- Slowly place 3 fish into the scene. Ask: *Are there 3 fish? Let's count them to check.* Count the fish together, pointing to each one. Click on the fish card to the right of the scene to display the numeral 3. Compare the numeral 3 on the screen with the 3 on the numeral card from PCM 5.
- Stand with your back to the children. Hold up the numeral card so they can see it. Slowly trace over the '3' with your finger. Say: *This is how we write the number 3. Now you write 3 in the air. Show me that again.*
- Click 'Clear all'. Click the fish card to hide the numeral 0.
- Repeat for other numbers up to 5.

Explore in groups

Counting collections station

- Spread numeral cards 1–5 from PCM 5 face up on the table. Next to each card place small pots. Ask the children to place small-world resources or counting apparatus into each pot to match the number on the card.

Beanbag hoops

- This activity would work well outdoors. Arrange 5 hoops (or similar) on the floor. Label them with numeral cards 1–5 from PCM 5 in random order. Invite children to put (or throw) the matching number of beanbags into each hoop. Prompt them to count the beanbags in each hoop to check that the quantity matches the numeral.

Pairs

- Provide sets of cards 1–5 from PCMs 1 (fruit) and 5 (numerals). Encourage children to play Pairs (see Generic games rules on page 220). Each pair or group will need one set of fruit cards and one set of numeral cards.
Variation: use cards 1–5 from PCMs 2, 3 or 4 instead of PCM 1.

Maths Foundation Activity Book A (adult-led)

Page 6 – Count and match

Teaching notes

Session 6: Write the numerals 1 to 5 (A)

Children begin to write the numerals 1 to 5.

You will need:

numeral cards 1–5 from PCM 5; various classroom objects and displays that have numbers; Unit 2 Session 6 slides; pencils; PCM 7; sand tray; paper; paint, water, aprons; *Maths Foundation Activity Book A*

Getting started

- Hold up the numeral card 3 from PCM 5. Ask children to say the number.
- Ask children to point to other examples of the number 3 in the classroom and/or outdoors.
- Repeat for the other numeral cards.

Teaching

- Display slide 1. Say: *We are going to learn how to write the number 1.*
- Use the row of numerals to demonstrate how to write the number 1. Slowly trace each numeral and explain how it is written. Write two extra number 1s at the end of the row.
- You could use the following rhyme: *Start at the top and down you run. That's how to write a one.*
- Ask children to write an imaginary 1 in the air with you, as you describe the formation of the numeral.
- Repeat for the numerals 2, 3, 4 and 5 using slides 2–5 in turn. You could use the following rhymes.
  - 2: *Around and back on a railroad track. Two, two! Two, two!*
  - 3: *Around the tree, around the tree. That's how to write a three.*
  - 4: *Down and over. Down once more. That's how to write a four.*
  - 5: *Down and around, then a line on high. That's how to write a five.*

Explore in groups

Practising numeral formation

- Give each child a pencil and a copy of PCM 7. Encourage them to practise writing the numerals 1 to 5.

Sand station

- Place a set of numeral cards 1–5 from PCM 5 near to the sand tray. Slightly dampen the sand and smooth it over. Let children take turns to write each of the numerals 1 to 5 in the sand.

Painting and drawing station

- Encourage children to finger-paint the numbers 1 to 5. Place a set of numeral cards 1–5 from PCM 5 on the table for children to refer to.

Maths Foundation Activity Book A (adult-led)

Page 7 – Trace and write

Session 7: Write the numerals 1 to 5 (B)

Children practise writing symbols (tally marks) and numerals to represent the numbers 1 to 5.

You will need:

numeral cards 1–5 from PCM 5; pencils; small pieces of paper; Unit 2 Session 7 slides; percussion instruments (optional); stapler; coloured pencils; PCMs 7 and 8; *Maths Foundation Reading Anthology*; *Maths Foundation Activity Book A*

Getting started

- Children sit at tables. Give each child a 1 to 5 numeral card from PCM 5.
- Display slide 1. Ask children to count the marks. Say: *If your card matches the number of marks, stand up.*
- Display slide 2 to confirm the answer. Ask the children to sit down.
- Repeat for the remaining slides.

Teaching

- Give each child a pencil and 5 small pieces of paper.
- Remind children that they have been learning how to record numbers: *We have used marks to keep a count. We have also learned how to write numbers from 1 to 5.*
- Explain that you will clap your hands (or play a percussion instrument) up to 5 times. Children must count the number of claps (sounds) and

make a mark for each count on one of their pieces of paper.
- Slowly clap 3 times.
- Ask children to hold up their piece of paper and say how many claps (sounds) you made.
- Now ask children to write 3 as a numeral on the back of that piece of paper.
- Ask children to hold up their piece of paper to show the numeral, and say the number.
- Repeat for each number 1 to 5.
- Staple together each child's pieces of paper in order, from 1 to 5. Label them with their name.

Explore in groups

Practising numeral formation

- Give each child a pencil and a copy of PCM 7 for additional practice writing the numerals 1 to 5.

Practising numeral formation and recognising cardinal values

- Before the activity, prepare copies of PCM 8. On each sheet, circle two numerals. Use a different coloured pencil for each circle, e.g. a blue circle around 3, and a red circle around 5. Give each child one of these sheets and two coloured pencils to match the colours of the circles. Ask children to trace each circled numeral in the corresponding colour. Then invite them to colour that number of balloons in the same colour (e.g. they colour 3 balloons blue and 5 balloons red).

Maths Foundation Reading Anthology (adult-led)

- Together, look at pages 8 and 9 (The stream) of the *Reading Anthology.* Give each child a small piece of paper. Tell each child (or point to) a different creature in the picture. Ask them to draw a small picture of their creature. Ask children to count how many of their creature they can see, and write the number as a numeral on their paper. If time allows, repeat. Ask children to compare their pieces of paper. Do all the pieces of paper with the same creature have the same numeral?

Maths Foundation Activity Book A (adult-led)

Pages 8 and 9 – Trace and circle

Note that there are 20 children on the two pages but only 15 will be circled.

Session 8: Compare two quantities and numbers to 5

Children use the words 'more' or 'fewer' to compare two quantities, and the words 'more' or 'less' to compare two numbers.

You will need:

cards 1–5 from PCMs 1 and 5; Unit 2 Session 8 slides; PCM 6; coloured pencils or crayons; 6 counters (or similar); stapler or glue; *Maths Foundation Activity Book A*; pencils

Getting started

- Hold up two of the cards (1–5 only) from PCM 1. Ask, for example: *Are there more apples or bananas?*
- Repeat several times. Hold up two cards each time. Ask questions such as: *Are there fewer … or …? Are there more … or …? Which card shows more fruit? Which card shows fewer fruit?*

Teaching

- Display slide 1. Ask: *How many red counters are there? How many yellow counters? Which is more: 2 or 4?*
- Point to the numbers 2 and 4 on the number track: *4 is more than 2.* Emphasise the word 'more'.
- Repeat for slides 2 to 5. Each time, use the number track to show the position of each number in the counting sequence: *… is more than …*
- Display slide 1 again. Ask: *How many red counters? How many yellow counters? There are fewer red counters than yellow counters.* Emphasise the word 'fewer'.
- Point to numbers 2 and 4 on the number track: *2 is less than 4.* Emphasise the word 'less'.
- Repeat for slides 2 to 5. For each slide, ask children to say which colour counters there are 'fewer' of. Use the number track to show the position of each number in the counting sequence, saying: *… is less than …*
- Hold up two of the numeral cards (1–5 only) from PCM 5. Ask, for example: *Which is more: 5 or 2?*

Teaching notes

- Repeat several times. Hold up two cards each time. Asking questions such as: *Which is less: … or …? Which is more: … or …?*

Explore in groups

Which tree has more?

- Before the activity, prepare copies of PCM 6. On each sheet, write a number from 1 to 5 above each tree. Invite children to draw the corresponding number of apples (or other fruit) on each tree. Then prompt them to draw a ribbon around the tree with *more* (or *fewer*) apples/fruit.

Comparing numbers (adult-led)

- Shuffle two sets of numeral cards 1–5 from PCM 5. Spread the 10 cards face down on the table. Ask two children to each turn over a card. Ask: *Who has the number that is more?* Give the child a counter (or similar). If two children turn over matching cards, and say that they are 'the same', give each child a counter. Continue until all 10 cards have been used.

Variation: ask: *Who has the number that is less?* Give the child a counter.

More or less cards

- Before the activity, prepare some double-sided cards. Use numeral cards 1–5 from PCM 5. Staple or glue pairs of cards back-to-back. You will need at least 3 or 4 double-sided cards for each child. Invite children to look at the number on each side of the card and circle the number that is *more*.

Variation: ask children to circle the number that is *less*.

Maths Foundation Activity Book A (adult-led)

Pages 10 and 11 – More

When children have completed both pages, ask them to point to the photo on each wall that shows *fewer* people.

Assessment opportunities

Assess children's learning against the objectives for this unit, using the guidance on formative assessment on pages 24–25, and record your observations in the Unit 2 progress tracking grid on page 27. The relevant pages of *Activity Book A* can also be used for assessment.

Can the children:

- show an understanding of the five key counting principles and of conservation of number?
- recite the number names 1 to 5 in order, counting forwards and backwards?
- count, with increased accuracy, up to 5 objects and things that cannot be touched?
- recognise the numerals 1 to 5?
- use marks and make attempts at writing the symbols (numerals) used to represent numbers 1 to 5?
- use the words 'more' or 'fewer' to compare two quantities from 1 to 5?
- use the words 'more' or 'less' to compare two numbers from 1 to 5?

Unit 3 Position, direction and movement

Theme: My senses

Overview

This unit introduces children to spatial awareness and spatial relationships. The most important aspect to develop in this unit is the vocabulary, with words constantly emphasised alongside everyday real-world experiences personal to the children. Opportunities should also be modelled and practised outside of the maths session as part of children's play and other general classroom experiences such as when tidying up.

During this unit, find opportunities throughout the day to use the *Number fluency* activities on pages 198–207, to continue to practise counting and recognising, reading, writing and comparing numbers to 5.

Learning objectives

Geometry and Measure – Position, direction and movement	• **7a Begin to understand and use the vocabulary of position, direction and movement.**

Learning objectives in bold are taught for the first time in this unit.

Vocabulary

position, above, below, on, next to, in, outside, in front, behind, direction, up, down, forwards, backwards, near, far, movement, turn, roll, slide

Making connections

English: My senses

Science: My senses

Preparation

You will need:

- *Maths Foundation Reading Anthology*, pages 2–9, 20 and 21.
- *Maths Foundation Activity Book A*, pages 10–15.
- PCMs 6 and 17.
- Magazines, comics, catalogues, toy packaging, etc.
- Selection of different small-world resources, e.g. people, animals and sealife, transport.
- Bear or similar soft toy.
- Box large enough for the soft toy to fit into.
- Paper, pencils, coloured pencils and/or crayons.
- Small pots or cups.
- 'u-d' counters (counters with 'u' for 'up' written on one side, and 'd' for 'down' on the other).
- 'f-b' counters (counters with 'f' for 'forwards' written on one side, and 'b' for 'backwards' on the other).
- 1 to 6 dot dice.

Teaching notes

- Counters.
- Interlocking cubes.
- Outdoor equipment and resources.
- Selection of small 'treasure' items, e.g. shiny stickers, costume jewellery, gold and silver chocolate coins.
- Box to be used as a treasure chest.
- Items that turn, e.g. hands of a clock, wheels, taps, keys in locks, screw-top lids on jars, nut and bolts.
- Items that slide, e.g. building blocks, boxes, books.
- Items that roll, e.g. balls, beads, marbles, oranges, plastic or wooden eggs.
- Items that slide and roll, e.g. tin cans, cotton reels, coins.
- Imaginative play station: Home corner area and props.
- Sand station: sand tray/pit, moulds, shells, pebbles, flags.
- Construction station: construction materials, e.g. building blocks, toy construction bricks, recycled boxes and packaging.

Before starting

As part of Session 1, start to create a wall and table display of images and objects associated with the vocabulary of position, direction and movement. Throughout the unit, ask children to look for related examples in magazines, comics, catalogues, toy packaging, etc. to include in the display. You could also include some suitable real-world examples.

Children will get the most out of this unit if they already have some experience of identifying their position relative to different objects and to others, using everyday words, and ways personal to them.

Common difficulties

Language is the main barrier to children demonstrating their understanding of spatial relationships. In English, there are many different terms used for position, direction and movement. Some are very similar, e.g. 'under' and 'below'. Show children repeated physical examples of spatial relationships, especially in everyday life experiences, and continually emphasise the key words.

Maths background

The Vocabulary section above lists the words that are focused on in this unit. However, you may wish to teach children some of the other, related vocabulary. For example:
- position words: over, under, underneath, top, bottom, side, beside, inside, around, front, back
- direction words: sideways, close.

Session 1: Use everyday language to describe position, direction and movement

Children use everyday language to describe position, direction and movement.

You will need:

selection of different small-world resources; Home corner area and props

Getting started

- Begin the unit by informally assessing children's understanding of some of the words associated with position, direction and movement.

Teaching

- Ask children to undertake specific tasks to demonstrate their understanding of the language of position, direction and movement. Each time make sure to emphasise the key word. For example:

  - *Everyone point to something that is **above**/**below** your nose.*

  - *Put this apron **on** a hook.*

- *Put this teddy **in**/**next to**/**outside** the hoop.*
- *Stand **behind**/**in front** of my desk.*
- *Stand **up**/sit **down**.*
- *Walk **forwards**/**backwards**.*
- *Stand **near**/**far** from the door.*
- ***Slide** that tray to me.*
- ***Roll** that ball to me.*
- ***Turn** in a circle.*

Explore in groups

Repeat *Teaching* (adult-led)

- Repeat the *Teaching* activity with smaller groups of children (outdoors, or in a different area, if appropriate). Occasionally, perform an action (e.g. stand in a hoop), and ask the children to describe what you have done.

Small-world play

- Let children play with small-world resources. Occasionally ask them to tell you about the position of objects, e.g. *What is in front of/ near/below/on the …?*

Imaginative play station

- When children are playing in the Home corner area, occasionally, ask them to tell you about the positions of objects, or themselves.

Session 2: Language of position (A)

Children use the words: 'above' and 'below'.

You will need:

Maths Foundation Reading Anthology; bear or similar soft toy; sand tray, moulds, shells, pebbles, flags; coloured pencils or crayons; PCM 6

Getting started

- During this session, focus on the words 'above' and 'below'.
- Introduce the bear.

Teaching

- Hold the bear above your head. Say: *Bear is above my head.* Emphasise the word 'above'.
- Place the bear below the table. Say: *Now Bear is below the table.* Emphasise the word 'below'.
- Move the bear to different positions. Each time, say whether the bear is *above* or *below* another object.
- Give the bear to a child. Ask them to place it in a particular position in the classroom, e.g. *Rafael, please put Bear below the sand tray.*
- Repeat several times, for different children and different positions.
- Show pages 20 and 21 (Aliya's room) of the *Reading Anthology*. Discuss the picture with the children. Point to examples that illustrate the words 'above' and 'below'. Ask individual children to come and point to objects in the picture that demonstrate their understanding of those key words.

Explore in groups

Maths Foundation Reading Anthology (adult-led)

- Together, look at pages 6 and 7 (Goldilocks and the three bears) of the *Reading Anthology*. Repeat the last part of the *Teaching* activity.

Sand station

- Provide various moulds and resources for children to use in the sand tray. When children are playing, occasionally ask them to tell you about the position of objects. Focus particularly on the words 'above' and 'below'.

Trees

- Provide copies of PCM 6. Invite children to draw on the picture (e.g. animals, birds, insects, fruit). Either give children instructions for where to draw ('above' or 'below'), or let children draw what they like and then ask questions about the positions.

Teaching notes

Session 3: Language of position (B)

Children use the words: 'on', 'in', 'outside' and 'next to'.

You will need:

Maths Foundation Reading Anthology; bear or similar soft toy; box large enough for the soft toy to fit into; outdoor equipment and resources; Home corner area and props

Getting started

- During this session, focus on the words 'on', 'in', 'outside' and 'next to'.
- Place the box on the floor next to a table.
- Put the bear on the table. Ask: *Where is Bear?* Discuss the children's responses.

Teaching

- Place the box on the table next to bear: *Bear is on the table.* Emphasise the word 'on'.
- Point to the box. Say: *Bear is also next to the box.* Emphasise the words 'next to'.
- Move the bear into different positions around the box. Each time, say where it is to reinforce the vocabulary.
- Give the bear to a child. Ask them to place it in a particular position in the classroom, e.g. *Mira, please put Bear in the box.*
- Repeat several times, with different children and different positions.
- Show pages 20 and 21 (Aliya's room) of the *Reading Anthology*. Discuss the picture with the children. Point to examples that illustrate the words 'on', 'in', 'outside' and 'next to'. Ask individual children to come and point to objects in the picture that demonstrate their understanding of those key words.

Explore in groups

Maths Foundation Reading Anthology
(adult-led)

- Together, look at pages 2 and 3 (At home) of the *Reading Anthology*. Repeat the last part of the *Teaching* activity.

Outdoor play

- Let children play freely with outdoor equipment and resources. Occasionally, ask them to describe the positions of objects or themselves, e.g. *Where are you standing? (In the hoop.) Where is Taha? (On a stepping stone.) Where is the ball? (Next to the sand tray.)*

Imaginative play station

- When children are playing in the Home corner area, ask them to place objects in particular positions, using the words 'on', 'in', 'outside' and 'next to'.

Session 4: Language of position (C)

Children use the words: 'in front' and 'behind'.

You will need:

Maths Foundation Reading Anthology; bear or similar soft toy; box large enough for the soft toy to fit into; construction materials, e.g. building blocks, toy construction bricks, recycled boxes and packaging; *Maths Foundation Activity Book A*; coloured pencils

Getting started

- During this session, focus on the words 'in front' and 'behind'.
- Ask a child to come and sit at the front. Place the bear on the floor in front of the child. Say: *Bear is in front of Pablo.* Place the bear behind the child. Say: *Now Bear is behind Pablo.*

Teaching

- Put the box on a table. Put bear in front of the box. Ask: *Where is Bear?* Discuss the children's responses.
- Say: *Bear is on the table. Bear is also in front of the box.* Emphasise the words 'in front'.
- Move the bear behind the box. Ask: *Where is Bear now?* Take responses, and emphasise that the bear is 'behind' the box.
- Give the bear to a child. Ask them to place it in a particular position in the classroom, e.g. *Lili, please put Bear in front of the bookcase.*
- Repeat several times, with different children and different positions.
- Show pages 20 and 21 (Aliya's room) of the *Reading Anthology*. Discuss the picture with the children. Point to examples that illustrate the words 'in front' and 'behind'. Ask individual children to come and point to objects in the

picture that demonstrate their understanding of those key words.

Explore in groups

Maths Foundation Reading Anthology (adult-led)

- Together, look at pages 4 and 5 (In the garden) of the *Reading Anthology*. Repeat the last part of the *Teaching* activity.

Construction station

- Provide construction materials. Let children create models. Occasionally, ask questions about the positions of parts of their model. Praise correct use of the positional language learned in the unit so far.

Maths Foundation Activity Book A (adult-led)

Page 12 – Position

Session 5: Language of direction

Children use the words: 'up', 'down', 'forwards' and 'backwards'.

You will need:

bear or similar soft toy; PCM 17; small pots or cups; 'u-d' counters (counters with 'u' for 'up' written on one side, and 'd' for 'down' on the other); 'f-b' counters (counters with 'f' for 'forwards' written on one side, and 'b' for 'backwards' on the other); 1 to 6 dot dice; counters; paper, coloured pencils or crayons

Getting started

- Children will need lots of space to move around. You may want to take them outdoors or to a larger room.
- Ask children to sit on the floor in front of you. Move yourself, and parts of your body, to introduce the directional words 'up', 'down', 'forwards' and 'backwards'. Describe the movements each time, using the key words.
- Then move the bear around (e.g. hold it up in the air, then down to the floor, then move it forwards, then backwards). Describe the movements, again emphasising the key words.

Teaching

- Ask children to spread out across the space, so that they do not bump into each other.
- Give instructions such as: *Walk forwards. Put your hands up. Crouch down. Jump up. Walk backwards. Look down.*

Explore in groups

Up or down

- Before the activity, prepare copies of PCM 17: cut out the three sections of the track and glue them together. Lay the tracks out *vertically*. Provide counters and small pots or cups each containing a 'u-d' counter and a 1 to 6 dot dice. Invite children to play a game in pairs. Each child places a counter on the handprint in the centre of the track. They take turns to tip the 'u-d' counter and dice from a pot, and move their counter to match. For example, rolling 'd' and '5' means move down 5 spaces. The winner is the first child whose counter comes off an end of the track.

Forwards or backwards

- Before the activity, prepare copies of PCM 17: cut out the three sections of the track and glue them together. Lay the tracks out *horizontally*. Provide counters and small pots or cups each containing a 'f-b' counter and a 1 to 6 dot dice. Invite children to play a game in pairs. Each child places a counter on the handprint in the centre of the track. They take turns to tip the 'f-b' counter and dice from a pot, and move their counter to match. For example, rolling 'f' and '4' means move forwards 4 spaces. The winner is the first child whose counter comes off an end of the track.

Draw for me

- Encourage children to play a drawing game in pairs. Give each child a sheet of paper and a pencil or crayon. One child holds the pencil/crayon on the paper. The other child tells them where to move it to draw a picture or pattern. Then ask children to swap roles and repeat.

Teaching notes

Session 6: Language of position and direction

Children follow and give instructions using the language of position and direction.

You will need:

outdoor equipment and resources; interlocking cubes; toy construction bricks; *Maths Foundation Activity Book A*; coloured pencils

Getting started

- Before the session, set out a selection of resources (any readily available classroom resources).
- Ask a child to place one of the resources in a particular position in the classroom. Use one of the positional words from previous sessions: 'above', 'below', 'on', 'in', 'outside', 'next to', 'in front' or 'behind'. For example: *Allison, put this pencil on my desk. Jaden, put this dice next to the box of cubes. Mariam, put this puppet below the table.*
- Repeat several times.

Teaching

- Give a child (or several children) an instruction to move either all of their body, or a body part. Use the directional words from the previous session: 'up', 'down', 'forwards' and 'backwards'. For example: *Amir, take two steps backwards. Tanvi, jump forwards. Bo, put your hands up.*
- Repeat several times.
- Choose a child who seems confident with the vocabulary. Give them one of the resources. Ask them to give instructions to another child to place the resource somewhere, e.g. *Lena, will you tell Deon where you would like him to put this plate?*
- Repeat several times. When needed, prompt children to use the positional and directional words learned so far.

Explore in groups

Outdoor play

- Let children play freely with outdoor equipment and resources. Occasionally, ask them to perform an action using positional and directional language. For example: *Ride the scooter forwards (or backwards). Walk below the frame. Stand on the stepping stone. Walk behind the water table.*

Tower of cubes

- Provide a tray of interlocking cubes. Invite children to play in pairs, building towers. Ask each child to take a cube of the same colour, e.g. blue. Then suggest that they take turns to give an instruction for what cube their partner should add to their tower next, e.g. *Put a red cube above the blue cube.* Encourage children to continue until each child has made a tower of at least 7 cubes. Ask children to compare their towers.

Construction station

- Provide identical sets of 8 toy construction bricks. Invite children to play a game in pairs. One child makes a model using their bricks, out of sight of their partner. That child then describes how they made their model. Their partner listens carefully and tries to make an identical model. When they have finished, they compare the two models. Children could then swap roles and repeat.

Maths Foundation Activity Book A (adult-led)

Page 13 – Direction

Session 7: Language of distance

Children use the words: 'near' and 'far'.

You will need:

Maths Foundation Reading Anthology; selection of small 'treasure' items such as shiny stickers, costume jewellery, gold and silver chocolate coins; box to be used as a treasure chest; selection of different small-world resources; outdoor equipment and resources; *Maths Foundation Activity Book A*; coloured pencils

Getting started

- Before the session, hide several small 'treasure items' around the room and outdoors.
- Give individual children instructions to find the position of one of the treasure items, using the positional and directional words learned so far. Tell them to bring it back to you to put into the 'treasure chest'.

Teaching

- Take one of the treasure items from the chest. Place it somewhere in the classroom visible to all the children.
- Give a child another treasure item. Ask them to place it near the other item. Emphasise the word 'near'.
- Then ask a child to place another treasure item far from the other items. Emphasise the word 'far'.
- Repeat several times, asking individual children to place treasure items near to or far from other objects in the classroom.
- Show pages 20 and 21 (Aliya's room) of the *Reading Anthology*. Discuss the picture with the children. Point to examples that illustrate the words 'near' and 'far'. Ask individual children to come and point to objects in the picture that demonstrate their understanding of those key words.

Explore in groups

Maths Foundation Reading Anthology
(adult-led)

- Together, look at pages 8 and 9 (The stream) of the *Reading Anthology*. Repeat the last part of the *Teaching* activity.

Small-world play

- Let children play with small-world resources. Occasionally, ask children to put a particular resource near to/far from another resource.

Outdoor play

- Let children play freely with outdoor equipment and resources. Occasionally, ask children to move or ride near to/far from different resources or children.

Maths Foundation Activity Book A (adult-led)

Page 14 – Near or far

Session 8: Language of movement

Children use the words: 'turn', 'slide' and 'roll'.

You will need:

Maths Foundation Reading Anthology; items that turn; items that slide; items that roll; items that slide and roll; *Maths Foundation Activity Book A*; coloured pencils

Getting started

- Explore and discuss items that turn, e.g. hands of a clock, wheels, taps, keys in locks, screw top lids on jars, nut and bolts.
- Demonstrate and explain to children how each item works, each time emphasising the word 'turn'.
- Demonstrate and explain how the word 'turn' can be used to describe a movement of self. For example, turn around on the spot, walk forwards then turn and face another direction and continue to walk forward. (At this stage, do not introduce the words 'left' and 'right'.)
- If space, ask children to walk about the room. Call out: *Stop! Now turn. Start walking again.* Repeat several times.

Teaching

- Show children one of the items that slides, e.g. a building block or a book. Demonstrate how it slides. Emphasise the word 'slide'. Repeat for another item that slides.
- Show children one of the items that rolls, e.g. a ball, orange, or a plastic or wooden egg. Demonstrate how it rolls. Emphasise the word 'roll'. Repeat for another item that rolls.

Teaching notes

- Show children one of the items that both slides and rolls, e.g. a tin can or a coin. Demonstrate how it both slides and rolls. Repeat for another item that slides and rolls.
- Ask individual children to go and find something in the classroom that either slides, rolls, or both slides and rolls. Ask the child to demonstrate the movement of their item to the rest of the children.

Explore in groups

Things that turn, slide and/or roll

- Allow children to freely explore the classroom, looking for, and testing out, objects that turn, slide, roll, and both roll and slide. Ask them to group the items according to the way they move.

Maths Foundation Reading Anthology
(adult-led)

- Together, look at pages 4 and 5 (In the garden) of the *Reading Anthology*. Discuss the picture with the children. Ask them to point to examples that illustrate the words: turn, slide and roll.

Maths Foundation Activity Book A (adult-led)

Page 15 – Slide or roll

Assessment opportunities

Assess children's learning against the objectives for this unit, using the guidance on formative assessment on pages 24–25, and record your observations in the Unit 3 progress tracking grid on page 27. The relevant pages of *Activity Book A* can also be used for assessment.

Can the children:
- use language to describe position, direction, and movement, including following and giving instructions?

Unit 4 Length and height

Theme: My pets

Overview

This unit introduces children to length, height and width.

The most important aspects of measures at this stage are for children to make direct and indirect comparisons of two objects, and to hear accurate vocabulary modelled.

Direct comparisons involve aligning two objects together to make them easy to compare. *Indirect comparison* is the process of comparing objects that are not aligned, such as the heights of a desk and a window.

Children will start describing and making comparisons using the language of size: big, bigger, small and smaller. They are swiftly moved on to more precise language. Terms such as long, tall, short, wide and narrow are used to describe length, height and width; terms such as longer, taller, shorter, wider and narrower are used to compare two objects.

Learning objectives

Geometry and Measure – Measurement	• **8a Use everyday language to describe and compare length, height and width, including long, longer, short, shorter, tall, taller, wide, wider, narrow and narrower.**

Learning objectives in bold are taught for the first time in this unit.

Vocabulary

measure, size, compare, same, big, bigger, small, smaller, length, height, long, longer, short, shorter, tall, taller, width, wide, wider, narrow, narrower

Making connections

English: Pets

Science: Animals

Preparation

You will need:

- *Maths Foundation Reading Anthology*, pages 10 and 11, 22 and 23.
- *Maths Foundation Activity Book A*, pages 16–19.
- Unit 4 slides (optional).
- PCMs 6 and 14.
- Rhymes and songs: *I stretch up tall* (21) (page 216).
- Selection of different objects suitable for describing and comparing sizes of two objects (including some objects that are the same length, height and width), e.g.

ribbon, string, fabric, wallpaper, shoes, socks, straws, playdough 'worms', pencils, crayons, paintbrushes, building blocks, rods of interlocking cubes, small-world resources, soft toys, recycled boxes and packaging.
- Tray.
- Interlocking cubes.
- Other objects suitable for sorting by size, e.g. buttons, beads, counters, natural materials (leaves, sticks, stones, pine cones).

Teaching notes

- Small sorting hoops.
- Sorting trays.
- Bags.
- Chalk.
- Magazines, newspaper, comics, catalogues, toy packaging, etc.
- Wallpaper or coloured paper.
- 1 to 6 dot dice.
- Painting and drawing station: paper, pencils, coloured pencils and crayons, chalk, paint, water, paintbrushes of different widths, aprons, sponges of various sizes.
- Imaginative play station: Home corner area, props (if possible, set this up as a vet's clinic).
- Playdough station: playdough.
- Construction station: construction materials, e.g. building blocks, toy construction bricks, recycled boxes and packaging, glue or tape.

Before starting

As part of Session 1, start to create a wall and table display of images and objects linked to length, height and width. Throughout the unit ask children to look for examples to include in the display (either real-world examples, or pictures from magazines, comics, catalogues, toy packaging, etc.).

Children will get the most out of this unit if they already have some experience of using everyday language to describe size, e.g. *That shirt is too big for me. That's a small chair.*

As part of each session, try to take children outdoors so that they can experience a range of different objects associated with length, height and width.

Common difficulties

We use a range of terms to describe size, which can be confusing for children. For example, 'short' is used to describe both length and height. Constantly model and reinforce the correct language.

Children will take some time to use these words accurately: 'bigger' (overall size), 'longer' (length) and 'taller' (height). Keep reinforcing the correct vocabulary, and provide plenty of visual representations to illustrate them.

Children will find it more difficult to compare sizes if the objects are not aligned: *indirect comparison*. Where possible, allow children to rearrange objects to make them easier to compare.

Maths background

We have two different ways of talking about 'height':

1) Tall/short: the distance from the bottom of an object to the top of it. For example: *An adult is tall. A child is short.*

2) High/low: the height of something from the ground. For example: *The child on the climbing frame is high up. The child crawling through the tunnel is low down.*

This has the potential to be confusing, particularly when children are first learning the vocabulary of height. For this reason, this unit focuses on tall/short, and does not introduce high/low.

Session 1: Describe sizes

Children begin to use language associated with size: 'big' and 'small'.

You will need:

selection of objects suitable for describing size ('big' and 'small'), but that focus on length and height, e.g. ribbon, string, socks, straws, playdough 'worms', pencils, rods of interlocking cubes, soft toys of different heights, recycled boxes and packaging; small sorting hoops; PCMs 6 and 14; paper; coloured pencils and crayons; paint, water, paintbrushes, aprons; sponges of various sizes

Getting started

- Before the session, place on a table a selection of objects suitable for describing size.
- Allow children to spend a few minutes looking at, and picking up, the objects.

Teaching

- Ask a child to stand beside you. Put your hand above your head, then above the child's head. Say: *I'm big. Yousaf is small.* Emphasise the words 'big' and 'small'.
- Hold up a small object, e.g. a pencil. Say: *This is small.* Point to a big object, e.g. a cupboard. Say: *This is big.*
- Hold up a small object, e.g. a button. Ask: *What can you tell me about the size of this object?* Emphasise the word 'size'. *Is it big or small?* Repeat for other objects.
- Ask children to make themselves as big as they can, then as small as they can. Repeat several times.
- Ask a small group of children to each find and hold up a small object. Ask the other children: *Are they right? Are these objects small?*
- Ask individual children to go and point to a big object. Ask the other children to say whether they are correct.

Explore in groups

Sorting by size

- Provide two sorting hoops and a selection of objects suitable for sorting by size (but that

focus on length and height). Invite children to sort the objects into two sets: 'big' and 'small'. Ask them to tell you how they sorted the objects.

Painting and drawing station

- Invite children to draw a big and a small pet of the same type, e.g. a big cat and a small cat.

Painting or drawing station

- Invite children to paint a set of big paw prints and a set of small paw prints. They could use either a brush or sponges.

Big and little spots (Big and little apples)

- Provide ladybird cards from PCM 14. Ask the child to draw big spots on one side of the ladybird and small spots on the other.

Variation: use PCM 6 instead. Ask children to draw big apples on one tree and small apples on the other.

Session 2: Compare sizes

Children compare the sizes of two objects. They use the words 'bigger' and 'smaller'.

You will need:

selection of objects suitable for comparing sizes ('bigger' and 'smaller'), but that focus on length and height, e.g. ribbon, string, fabric, socks, straws, playdough 'worms', pencils, towers of interlocking cubes, soft toys of different heights, recycled boxes and packaging; trays or bags; Home corner area; *Maths Foundation Reading Anthology*; *Maths Foundation Activity Book A*; pencils

Getting started

- Before the session, place on a table a selection of objects suitable for comparing sizes.
- Remind children of the words 'big' and 'small'.
- Point to, or hold up, different objects. Ask: *What can you tell me about the size of this object? Is it big or small?* Emphasise the words 'size', 'big' and 'small'.

Teaching notes

- Ask children to make themselves as big as they can, then as small as they can. Repeat several times.

Teaching

- Hold up two similar objects of different lengths/heights, e.g. two straws. Say: *This straw is bigger than this straw.* Emphasise the word 'bigger'.
- Then hold up two soft toys of different sizes, e.g. bears. Say: *This bear is smaller than this bear.* Emphasise the word 'smaller'.
- Repeat for other pairs of objects, e.g. two clay 'worms', two towers of interlocking cubes.
- Hold up two *different* objects of different lengths/heights, e.g. a crayon and a paintbrush. Say: *This paintbrush is bigger than this crayon. The crayon is smaller than the paintbrush.*
- Repeat for other pairs of different objects, e.g. a length of string and a length of ribbon, a tower of interlocking cubes and a tower of building blocks.
- Hold up, or point to, an object, e.g. a sock. Ask individual children to go and point to something that is bigger than/smaller than your object. Repeat for different objects.

Explore in groups

'Smaller' hunt

- Provide bags, and a variety of objects (e.g. soft toy, building block, small recycled box, sock). Invite children to take an object and a bag, and go on a size hunt around the room (and outdoors, if possible). Ask them to collect objects that are smaller than their object.

Variation: ask children to look for objects that are bigger than their object.

Imaginative play station

- If possible, set up the Home corner area as a vet's clinic. Provide lots of different toy animals. You could suggest that children have a 'small pets' area and a 'big pets' area. Ask: *How will you decide which animals to put in each area? Which of these two animals is bigger?*

Maths Foundation Reading Anthology
(adult-led)

- Together, look at pages 10 and 11 (How many animals?) of the *Reading Anthology*. Discuss and compare the sizes of the different animals in the pictures.

Maths Foundation Activity Book A (adult-led)

Page 16 – Bigger

Session 3: Describe lengths

Children describe the length of objects. They use the words 'long' and 'short'.

You will need:

Maths Foundation Reading Anthology; selection of objects suitable for describing length, e.g. ribbon, string, fabric, socks, straws, playdough 'worms', pencils, rods of interlocking cubes, recycled boxes and packaging; small sorting hoops; paper; paint, water, aprons; crayons, chalk

Getting started

- Before the session, place on a table a selection of objects suitable for describing length. Position the objects horizontally, rather than vertically.
- Allow children to spend a few minutes looking at, and picking up, the objects.

Teaching

- Point to, or hold up, different objects. Say, for example: *This whiteboard is long. This pencil is short. This is a long piece of string. This is a short piece of ribbon. This is a short worm, and this is a long worm.* Emphasise the words 'long' and 'short'.
- Point to, or hold up, more objects. Ask: *What can you tell me about the length of this straw?* Emphasise the word 'length'. Explain: *Length tells us how long something is from one end to the other.*
- Ask children to use their hands to show you what long means, then to show you what short means.
- Ask individual children to go and point to something that is long/short.
- Show pages 22 and 23 (At the park) of the *Reading Anthology*. Discuss the picture. Ask children to describe what they can see, using the words 'long' and 'short'.

Explore in groups

Sorting by length

- Provide two sorting hoops and a selection of objects suitable for sorting by length. Invite children to sort the objects into two sets: 'long' and 'short'. Ask them to tell you how they sorted the objects.

Painting and drawing station

- Invite children to paint a long snake and a short snake, using a finger dipped in paint.

Crayon walks

- Give each child a sheet of paper and two different coloured crayons. Ask children to take each crayon on a 'walk' around the paper. They take one crayon on long 'walks' (drawing long lines), and the other crayon on short 'walks' (drawing short lines).

Variation: ask children to use chalk instead of crayons, outdoors.

Session 4: Compare lengths

Children compare the lengths of two objects. They use the words 'longer', 'shorter' and 'the same length'.

You will need:

Maths Foundation Reading Anthology; selection of objects suitable for comparing lengths (include some objects that are the same length), e.g. ribbon, string, fabric, socks, straws, playdough 'worms', pencils, rods of interlocking cubes, recycled boxes and packaging; bags; playdough; *Maths Foundation Activity Book A*; pencils

Getting started

- Before the session, place on a table a selection of objects suitable for comparing lengths. Position the objects horizontally, rather than vertically.
- Remind children of the words 'long' and 'short'. Ask a child to come and stand beside you. Take a long step, leaving both feet apart. Ask the child to take a step. Say: *I took a long step. Amina took a short step.* Emphasise the words 'long' and 'short'.

- Ask children to use their hands to show you what long means and then to show you what short means.
- Ask individual children to go and point to a long/short object.

Teaching

- Take two similar objects of different lengths (e.g. two pieces of string). Demonstrate how to line them up carefully to make their lengths easy to compare: *direct comparison*. Point and say: *This piece of string is longer than this piece of string. This piece of string is shorter than this piece of string.* Emphasise the words 'longer' and 'shorter'.
- Ask a child to compare two other similar objects, e.g. two playdough 'worms'. Help them to align them carefully. Ask them to say which is longer, and which is shorter. Ask the rest of the children if they agree.
- Next directly compare lengths of two *different* objects, e.g. a crayon and a paintbrush. Point and say: *This crayon is shorter than the paintbrush. The paintbrush is longer than the crayon.*
- Ask a child to compare two other different objects, e.g. a rod of interlocking cubes and a pencil. Help them to align them carefully, then say which is longer.
- Then directly compare two objects that are the same length. Point and say: *These two crayons are the same length.* Emphasise the phrase: 'the same length'.
- Now ask children to compare the lengths of two objects that can't be aligned together: *indirect comparison*. Point and say: *Which do you think is longer: the length of my desk or the length of this bookcase?*
- Point to, or hold up, an object, e.g. a straw. Ask individual children to go and point to something that is longer than/shorter than/the same length as your object.
- Show pages 22 and 23 (At the park) of the *Reading Anthology*. Ask children to compare objects in the picture using the words 'long', 'longer', 'short' and 'shorter'.

Explore in groups

'Shorter' hunt

- Provide bags, and a variety of objects (e.g. pencil, crayon, paintbrush, rod of interlocking cubes). Invite children to take an object and a bag, and go on a 'length' hunt around the room (and outdoors, if possible). Ask them to collect objects that are shorter than their object.

Variation: ask children to look for objects that are longer than their object.

Playdough station

- Let children play with playdough. Encourage them to roll pieces of playdough to make 'snakes'. Ask them to tell you which of two snakes is longer.

Maths Foundation Reading Anthology (adult-led)

- Together, look at pages 22 and 23 (At the park) of the *Reading Anthology*. Ask children to compare objects in the picture, using the words 'long', 'longer', 'short' and 'shorter'.

Maths Foundation Activity Book A (adult-led)

Page 17 – Longer

Session 5: Describe heights

Children describe the height of objects. They use the words 'tall' and 'short'.

You will need:

Maths Foundation Reading Anthology; selection of objects suitable for describing height, e.g. building blocks, towers of interlocking cubes, soft toys of different heights, recycled boxes and packaging; small sorting hoops; paper; coloured pencils and crayons; glue or tape; paint, water, paintbrushes, aprons

Getting started

- Before the session, place on a table a selection of objects suitable for describing height. Position the objects vertically, rather than horizontally.
- Say the action rhyme: *I stretch up tall* (21).

Teaching

- Ask a child to come and stand beside you. Putting your hand above your head, then

above the child's head say: *I'm tall. Zara is short.* Emphasise the words 'tall' and 'short'.
- Point to, or hold up, different objects. Say, for example: *This door is tall. Bear is short. This is a tall cupboard. This is a short tower of cubes. This is a tall tower of cubes.*
- Point to, or hold up, more objects. Ask: *What can you tell me about the height of this broom?* Emphasise the word 'height'. Explain: *Height tells us how far it is from the bottom of something to the top of it.*
- Ask children to use their hands to show you what tall means, then to show you what short means.
- Ask individual children to go and point to something that is tall/short.
- Show pages 22 and 23 (At the park) of the *Reading Anthology*. Ask children to describe what they can see, using the words 'tall' and 'short'.

Explore in groups

Sorting by height

- Provide two sorting hoops and a selection of objects suitable for sorting by height. Invite children to sort the objects into two sets: 'tall' and 'short'. Ask them to tell you how they sorted the objects.

Painting and drawing station

- Invite children to draw a tall animal (or pet) and a short animal (or pet).

Construction station

- Provide building blocks and/or recycled boxes and packaging. Let children build models. Ask questions about the height of their model. If appropriate, allow children to glue/tape their resources together and paint/decorate their model.

Variation: provide interlocking cubes. Ask children to build a tall tower and a short tower.

Session 6: Compare heights

Children compare the heights of two objects. They use the words 'taller', 'shorter' and 'the same height'.

You will need:

Maths Foundation Reading Anthology; selection of objects suitable for comparing heights (include some objects that are the same height), e.g. building blocks, towers of interlocking cubes, soft toys of different heights, recycled boxes and packaging; newspaper; tape; 1 to 6 dot dice; tray of extra interlocking cubes; *Maths Foundation Activity Book A*; pencils

Getting started

- Before the session, place on a table a selection of objects suitable for comparing heights. Position the objects vertically, rather than horizontally.
- Remind children of the words 'tall' and 'short'.
- Point to objects and furniture around you. Say, for example: *This whiteboard is tall. This box is short.*
- Ask children to use their hands to show you what tall means, then to show you what short means.
- Ask individual children to go and point to a tall/short object.

Teaching

- Take two similar objects of different heights (e.g. two towers of interlocking cubes). Demonstrate how to line them up carefully to make their heights easy to compare: *direct comparison.* Point and say: *This tower is taller than this tower. This tower is shorter than this tower.* Emphasise the words 'taller' and 'shorter'.
- Ask a child to compare two other similar objects, e.g. two soft toys. Help them to align them carefully. Ask them to say which is taller, and which is shorter. Ask the rest of the children if they agree.
- Next directly compare heights of two *different* objects, e.g. a box placed on the floor next to a cupboard. Point and say: *The cupboard is taller than the box. The box is shorter than the cupboard.*
- Ask a child to compare two other different objects, e.g. a door and the child themselves. Ask them to go and stand by the door, then say which is taller.
- Then directly compare two objects that are the same height. Point and say: *These two chairs*

are the same height. Emphasise the phrase 'the same height'.
- Now ask children to compare the heights of two objects that can't be aligned together: *indirect comparison.* Point and say: *Which do you think is taller: the height of my desk or the height of this cupboard?*
- Point to, or hold up, an object, e.g. a pencil. Ask individual children to go and point to something that is taller than/shorter than/the same height as your object.
- Show pages 22 and 23 (At the park) of the *Reading Anthology.* Ask children to compare objects in the picture using the words 'tall', 'taller', 'short' and 'shorter'.

Explore in groups

Taller and shorter towers

- Provide a selection of objects (e.g. tower of interlocking cubes, soft toy, recycled box, paintbrush), and lots of building blocks. Invite children to choose an object, and make two towers of building blocks: one that is taller than their object, and one that is shorter.

Paper towers

- Before the activity, roll up a sheet of newspaper and tape the edge to create a simple 'tower'. Provide sheets of newspaper and tape. Invite children to try to create a taller tower. Offer assistance where necessary.

Variation: ask children to create a shorter tower.

Comparing towers (adult-led)

- Provide a tray of interlocking cubes and a 1 to 6 dot dice. Ask each child to roll the dice, say the number, count out that number of cubes and use them to build a tower. Repeat, with each child rolling the dice again and adding cubes to their existing tower. Ask questions comparing the heights of two towers: *Whose tower is taller than Julia's? Whose tower is shorter than Kimi's? Does anyone have a tower the same height as someone else's?*

Maths Foundation Activity Book A (adult-led)

Page 18 – Taller

Teaching notes

Session 7: Describe widths

Children describe the width of objects. They use the words 'wide' and 'narrow'.

You will need:

Unit 4 Session 7 slide (optional); *Maths Foundation Reading Anthology*; selection of objects suitable for describing width, e.g. ribbon, fabric, wallpaper, shoes, paintbrushes, Home corner props, recycled boxes and packaging, natural materials (leaves, sticks, pine cones); small sorting hoops; paper (ideally green); pencils; scissors; paint, water, paintbrushes of different widths, aprons

Getting started

- Before the session, place on a table a selection of objects suitable for describing width. Finding objects suitable for teaching width may be difficult. If not enough real-world objects can be found, a slide is available.
- Begin by pointing to, or holding up, something long (e.g. a ribbon). Say: *Remember, 'length' tells us how long something is from one end to the other.*
- Next show children a chair. Indicate its height using your hands. Say: *Remember, 'height' tells us how far it is from the bottom of something to the top of it.*
- Then use your hands to indicate the chair's width. Say: *Today we will learn about 'width'. Width tells us how far it is from one side to the other.* Emphasise the word 'width'.

Teaching

- Point to a wide object (e.g. a window). Use your hands to indicate the width, and say: *This window is wide.* Point to a narrow object (e.g. a pencil). Use your fingers to indicate the width, and say: *This pencil is narrow.* Emphasise the words 'wide' and 'narrow'.
- Point to different objects and say, for example: *This door is wide. This cube tower is narrow. This is a wide rug. This is a narrow paintbrush.*
- Point to, or hold up, more objects. Ask children to say whether each one is wide or narrow.
- Point to, or hold up, more objects. Ask: *What can you tell me about the width of this object?*

- Ask children to use their hands to show you what wide means, then to show you what narrow means.
- Ask individual children to go and point to something that is wide/narrow.
- Show pages 22 and 23 (At the park) of the *Reading Anthology*. Ask children to describe what they can see, using the words 'wide' and 'narrow.

Explore in groups

Sorting by width

- Provide two sorting hoops and a selection of objects suitable for sorting by width. Invite children to sort the objects into two sets: 'wide' and 'narrow'. Ask them to tell you how they sorted the objects.

Wide and narrow leaves (adult-led)

- Provide paper (ideally green), pencils and scissors. Ask children to draw and cut out leaves. Encourage them to make one wide leaf and one narrow leaf.

Variation: take children outdoors and ask them to find wide and narrow leaves.

Painting and drawing station

- Provide paper, paint and paintbrushes of different widths. Invite children to use one colour of paint to make wide lines, and another colour to make narrow lines.

Session 8: Compare widths

Children compare the widths of objects. They use the words 'wider', 'narrower' and 'the same width'.

You will need:

Unit 4 Session 8 slides (optional); *Maths Foundation Reading Anthology*; selection of different objects suitable for comparing widths of two objects (include some objects that are the same width), e.g. ribbon, fabric, wallpaper, shoes, paintbrushes, soft toys, Home corner props, recycled boxes and packaging, natural materials (leaves, sticks, pine cones); paper; scissors; construction materials; *Maths Foundation Activity Book A*; pencils

Getting started

- Before the session, place on a table a selection of objects suitable for comparing widths. Finding objects suitable for teaching width may be difficult. If not enough real-world objects can be found, a set of slides is available.
- Remind children of the words 'wide' and 'narrow'.
- Point to objects and furniture around you. Say, for example: *The whiteboard is wide. My finger is narrow.*
- Ask children to use their hands to show you what wide means, then to show you what narrow means.
- Ask individual children to go and point to something that is wide/narrow.

Teaching

- Place two similar objects of different widths (e.g. two soft toys) beside each other. Say, for example: *The elephant is wider than the giraffe. The giraffe is narrower than the elephant.* Emphasise the words 'wider' and 'narrower'.
- Ask a child to compare two other similar objects, e.g. two paintbrushes of different widths. Ask them to say which is wider, and which is narrower. Ask the rest of the children if they agree.
- Next compare widths of two *different* objects, e.g. a cardboard box and a shoe. Point and say: *The box is wider than the shoe. The shoe is narrower than the box.*
- Ask a child to compare two other different objects, e.g. a door and the child themselves. Help them to go and stand by the door, then say which is wider.
- Then directly compare two objects that are the same width. Point and say: *These two chairs are the same width.* Emphasise the phrase 'the same width'.
- Now ask children to compare the widths of two objects that can't be aligned together: *indirect comparison.* Point and say: *Which do you think is wider: my desk or that poster?*
- Hold up, or point to, an object. Ask individual children to go and point to something that is wider or narrower than your object.
- Show pages 22 and 23 (At the park) of the *Reading Anthology*. Ask children to compare objects in the picture using the words 'wider', and 'narrower'.

Explore in groups

Paper strips (adult-led)

- Provide paper and scissors. Ask each child to cut a strip of paper. While they do this, cut two strips yourself, of equal width. Then ask children to compare the strips. Ask questions such as: *Whose strip is wider than Oliver's? Rowan, is your strip wider or narrower than this one? Are my two strips the same width?* Place one of your strips on the table. Ask children to each cut a strip wider (or shorter) than yours.

Construction station

- Provide construction materials (building blocks, toy construction bricks or recycled boxes and packaging). Let children build models. Ask questions about the width of their models, e.g. *Is your model wider or narrower than Luc's? How could you make your model wider than it is now?*

Maths Foundation Activity Book A (adult-led)

Page 19 – Wider

Assessment opportunities

Assess children's learning against the objectives for this unit, using the guidance on formative assessment on pages 24–25, and record your observations in the Unit 4 progress tracking grid on page 28. The relevant pages of *Activity Book A* can also be used for assessment.

Can the children:

- describe and compare the sizes of objects, using the words: 'big', 'bigger', 'small' and 'smaller'?
- describe and compare the lengths of objects, using the words: 'long', 'longer', 'short', 'shorter' and 'the same length'?
- describe and compare the heights of objects, using the words: 'tall', 'taller', 'short', 'shorter' and 'the same height'?
- describe and compare the widths of objects, using the words: 'wide', 'wider', 'narrow', 'narrower' and 'the same width'?

Unit 5 2D shapes

Theme: Plants

Overview

In this unit, children explore shapes in the world around them. They are introduced to the 2D ('flat') shapes: circle, triangle, square and rectangle. They learn to name 2D shapes and recognise them in different sizes and orientations. They compare and sort 2D shapes and begin to use simple shape vocabulary to describe them.

Children are also introduced to repeating patterns. Seeking and exploring patterns is at the heart of mathematics. (This is developed further in Unit 9: Patterns and data.) In this unit, children use colour and 2D shapes to observe, copy, continue and create simple 'AB' repeating patterns, such as square, triangle, square, triangle. They begin to make generalisations.

During this unit, find opportunities throughout the day to use the *Number fluency* activities on pages 198–207, to continue to practise counting and recognising, reading, writing and comparing numbers to 5.

Learning objectives

Geometry and Measure – Understanding shape	• **6a Identify, describe, compare and sort 2D shapes.**
Number – Patterns and sequences	• **5a Talk about, recognise and make simple patterns using concrete materials or pictorial representations.**
Statistics	• **10a Sort, represent and describe data using concrete materials or pictorial representations.**

Learning objectives in bold are taught for the first time in this unit.

Vocabulary

2D (2-dimensional), shape, flat, curved, straight, round, corner, side, sort, circle, triangle, square, rectangle, pattern, repeating pattern

Making connections

English: Plants

Science: Plants

Preparation

You will need:

- *Maths Foundation Reading Anthology*, pages 2–5, 16 and 17.
- *Maths Foundation Activity Book A*, pages 20–23.
- Digital Tool: Shape set.
- PCMs 18–23.

- Rhymes and songs: *Draw a circle* (22), *Draw a triangle* (23), *Draw a square* (24), *Draw a rectangle* (25) (pages 216–217).
- Magazines, comics, catalogues, toy packaging, etc.
- Selection of everyday objects with faces or parts that are circle-, triangle-, square- or rectangle-shaped.

- Selection of different small-world resources, e.g. people, animals and sealife, transport.
- Selection of different counting apparatus, e.g. cubes, counters.
- Sets of 2D geometric shapes.
- Non-transparent ('feely') bags.
- Sorting trays.
- Sorting hoops.
- Labels: 'circles', 'not circles', 'triangles', 'not triangles', 'squares', 'not squares', 'rectangles' and 'not rectangles'.
- Large books.
- Resources for making 2D shape patterns (circles, triangles, squares and rectangles), e.g. shape stickers; self-inking shape stamps; 3D geometric plastic cylinder, pyramid, cube, cuboid; paints or stamp pads.
- Painting and drawing station: paper, coloured pencils or crayons, paint, water, paintbrushes, aprons.

- Playdough station: playdough, readily available classroom resources (including some with circular faces).
- Sand station: sand tray/pit, 2D geometric shapes, dishes.

Before starting

As part of Session 1, start to create a wall display of 2D shapes. Throughout the unit ask children to look for examples of circles, triangles, squares and rectangles in magazines, comics, catalogues, etc. to include in the display.

Children will get the most out of this unit if they already have some experience of sorting familiar objects using criteria such as size and colour. They should be able to use everyday words to explain choices, including 'not'.

Common difficulties

Children often think that 2D shapes are always regular (all sides and angles equal) with horizontal bases. This can mean they fail to recognise shapes in different orientations (e.g. when a square has been rotated 45°), or irregular shapes (e.g. a scalene triangle). Show children a wide variety of regular and irregular 2D shapes in different orientations and sizes, emphasising the similarities and differences.

Children often find it difficult to distinguish between a square and a rectangle. Mathematically, a square *is* a rectangle; it's a special kind of rectangle that has equal sides. However, at this stage we want children to be able to see the difference between squares and rectangles. Provide lots of opportunities for them to feel, draw, and draw around both types of shape.

Session 1: Recognise 2D shapes in everyday life

Children begin to recognise and name 2D shapes in everyday life.

You will need:

selection of everyday objects with faces or parts that are circle-, triangle-, square- or rectangle-shaped; paper; coloured pencils or crayons, or paint, water, paintbrushes, aprons; selection of different small-world resources and counting apparatus; 2D geometric shapes; sorting trays; *Maths Foundation Reading Anthology*

Getting started

- Begin by pointing out examples of different real-world 2D shapes within the classroom. Point to different shapes in turn, saying the name of the shape. Focus on circles, triangles, squares and rectangles. Include a selection of regular and irregular shapes, and shapes of different sizes and in different orientations.
- It is important to clearly demonstrate to children the 2D faces on 3D shapes, so that they do not confuse 2D 'flat' shapes and 3D 'solid' shapes. For example, when pointing

Teaching notes

out the circular face of a clock, trace all the way round the face of the clock with your finger, to show the 2D ('flat') circle shape and not the 3D ('solid') cylinder shape. Similarly, when pointing to the square or rectangular face of a box, focus on the 2D shape and not the 3D cube or cuboid shape.

- Occasionally, ask children to describe a shape using their own words. (At this stage, do not introduce children to the *properties* of circles, triangles, squares and rectangles.) Ask them to say what is the same and what is different about various 2D shapes. Encourage them to recognise and name the 2D shapes and describe them as '2D' or 'flat' shapes.

Teaching

- Take children on a 2D shape hunt around the classroom and outdoors.
- Ask children to point to and name different 2D shapes. Where appropriate, encourage them to trace with their finger around different shapes, e.g. the rectangular face of a brick; squares or rectangles on a hopscotch game; shapes on gym equipment; circles on wheels/ tyres.
- Once children begin to become familiar with circles, triangles, squares and rectangles, arrange them into four groups: a circle group, a triangle group, a square group and a rectangle group. Ask all the members of each group to find, or point to, an example of their group's shape. Reinforce good examples of each shape, e.g. *Well done Hassan, that's a square. We can say that a square is a flat shape. We can also say that a square is a 2D shape.*
- Repeat several times, changing each group's shape.

Explore in groups

Painting and drawing station

- Invite children to draw or paint a picture based on their 2D shape hunt. Ask children to name some of the shapes in their pictures.

Sorting

- Provide sorting trays, and a selection of different small-world resources, counting apparatus and 2D geometric shapes. Encourage children to sort the resources using their own criteria. Ask them to explain how they have sorted. If they have not used shape

as criteria, ask them to consider sorting by shape. Can they name any of the 2D shapes?

Maths Foundation Reading Anthology (adult-led)

- Together, look at pages 2 and 3 (At home) of the *Reading Anthology*. Discuss the picture with the children. Ask them to point to and name examples of circles, triangles, squares and rectangles. Repeat for pages 4 and 5 (In the garden).

Session 2: Recognise and name circles

Children begin to recognise and name circles, including in different sizes.

You will need:

Shape set Digital Tool; 2D geometric shapes; non-transparent ('feely') bag; two sorting hoops; labels: 'circles' and 'not circles'; *Maths Foundation Reading Anthology*; paper; coloured pencils or crayons; playdough; small, readily available classroom resources (including some with circular faces); PCMs 18–22

Getting started

- Before the session, place some 2D geometric shapes in the 'feely' bag (include more circles than any other shape).
- Show children a circle. Briefly discuss its properties: a 2D, flat, round shape with one curved side. At this stage, do not expect children to recall these properties. They should just be able to recognise and name circles, and distinguish between shapes that are circles and shapes that are *not* circles.
- Rotate the circle to show that it looks the same in any orientation.
- Show other circles of different sizes. Constantly reinforce the words 'circle', '2D' and 'flat' so that children begin to recognise and classify circles as 2D, flat shapes.

Teaching

- Show children a shape that is *not* a circle, such as a square. Ask: *Is this shape a circle? What's the same about this shape and a circle? What's different about them?*
- Repeat for other 2D shapes that are *not* circles.

- Display the Shape set Digital Tool. Add two sets: one set labelled 'circles', the other set labelled 'not circles'.
- Place different 2D shapes into the correct sets. Click on individual shapes, especially circles, to change their colour and size.
- Ask individual children to come and place more shapes into the correct sets: *Jack, come and put another circle into the 'circles' set. Farida, come and put another shape into the 'not circles' set.*
- Place two sorting hoops on the floor. Label them 'circles' and 'not circles'. Ask a child to take a shape from the 'feely' bag. Ask: *Is this shape a circle?* Ask the child to place the shape into the correct hoop.
- Repeat several times.
- Teach children the action rhyme *Draw a circle* (22).

Explore in groups

Maths Foundation Reading Anthology
(adult-led)

- Together, look at pages 16 and 17 (2D shapes) of the *Reading Anthology*. Discuss the examples of circles in the picture. Ask children to draw some things that are circle-shaped. They can use ideas from the *Anthology*, or their own ideas.

Playdough station

- Provide a selection of small, readily available classroom resources (e.g. glue stick, crayon, counter, interlocking cube, dice, ruler, paintbrush, bead). Let children explore pressing different resources into playdough to see what shapes they can make. Ask them to tell you about any circle shapes they have made.

Circles and not circles

- Provide a set of circle, triangle, square, and rectangle cards from PCMs 18–21. Invite children to sort the cards into two sets: 'circles' and 'not circles'.

Variation: use 2D geometric shapes instead of the cards.

Find the circles

- Provide copies of PCM 22. Encourage children to colour all the circles red. When finished, you could ask children to draw three different circles on the back of their sheet.

Session 3: Recognise and name triangles

Children begin to recognise and name triangles, including in different sizes and orientations.

You will need:

Shape set Digital Tool; 2D geometric shapes; non-transparent ('feely') bag; two sorting hoops; labels: 'triangles' and 'not triangles'; *Maths Foundation Reading Anthology*; paper; coloured pencils or crayons; PCMs 18–21; sand tray; *Maths Foundation Activity Book A*

Getting started

- Before the session, place some 2D geometric shapes in the 'feely' bag (include more triangles than any other shape).
- Show children a triangle. Briefly discuss its properties: a 2D, flat shape with three straight sides. At this stage, do not expect children to recall these properties. They should just be able to recognise and name triangles, and distinguish between shapes that are triangles and shapes that are *not* triangles.
- Rotate the triangle to show it in different orientations.
- Show other triangles of different sizes and shapes (equilateral, isosceles, right-angled and scalene). Constantly reinforce the words 'triangle', '2D' and 'flat' so that children begin to recognise and classify triangles as 2D, flat shapes (including regular and irregular triangles, and triangles in different orientations).

Teaching

- Show children a shape that is *not* a triangle, such as a square. Ask: *Is this shape a triangle? What's the same about this shape and a triangle? What's different about them?*
- Repeat for other 2D shapes that are *not* triangles.
- Display the Shape set Digital Tool. Add two sets: one set labelled 'triangles', the other set labelled 'not triangles'.
- Place different 2D shapes into the correct sets. Click on individual shapes, especially triangles, to change their colour, size and orientation.
- Ask individual children to come and place more shapes into the correct sets.
- Place two sorting hoops on the floor. Label them 'triangles' and 'not triangles'. Ask a child

to take a shape from the 'feely' bag. Ask: *Is this shape a triangle?* Ask the child to place the shape into the correct hoop.
- Repeat several times.
- Teach children the action rhyme *Draw a triangle* (23).

Explore in groups

Maths Foundation Reading Anthology
(adult-led)

- Together, look at pages 16 and 17 (2D shapes) of the *Reading Anthology*. Discuss the examples of triangles in the picture. Ask children to draw some things that are triangle-shaped. They can use ideas from the *Anthology*, or their own ideas.

Triangles and not triangles

- Provide a set of circle, triangle, square, and rectangle cards from PCMs 18–21. Invite children to sort the cards into two sets: 'triangles' and 'not triangles'.

Variation: use 2D geometric shapes instead of the cards.

Sand station

- Before the activity, bury some 2D geometric shapes in the sand. Include more triangles than any other shape. Include different sizes and types of triangle. Let children hunt for the shapes. Ask them to tell you about the shapes they have found. *Which shapes are triangles? How do you know?*

Maths Foundation Activity Book A (adult-led)

Page 20 – Circles and triangles

Session 4: Recognise and name squares

Children begin to recognise and name squares, including in different sizes and orientations.

You will need:

Shape set Digital Tool; 2D geometric shapes; non-transparent ('feely') bag; two sorting hoops; labels: 'squares' and 'not squares'; *Maths Foundation Reading Anthology*; paper; coloured pencils or crayons; PCMs 18–22

Getting started

- Before the session, place some 2D geometric shapes in the 'feely' bag (include more squares than any other shape).
- Show children a square. Briefly discuss its properties: a 2D, flat shape with four straight sides all the same size. At this stage, do not expect children to recall these properties. They should just be able to recognise and name squares, and distinguish between shapes that are squares and shapes that are *not* squares, especially rectangles.
- Rotate the square to show it in different orientations.
- Show other squares of different sizes. Constantly reinforce the words 'square', '2D' and 'flat' so that children begin to recognise and classify squares as 2D, flat shapes (including squares in different orientations.)

Teaching

- Show children a shape that is *not* a square, such as a triangle. Ask: *Is this shape a square? What's the same about this shape and a square? What's different about them?*
- Repeat for other 2D shapes that are *not* squares.
- Display the Shape set Digital Tool. Add two sets: one set labelled 'squares', the other set labelled 'not squares'.
- Place different 2D shapes into the correct sets. Click on individual shapes, especially squares, to change their colour, size and orientation.
- Ask individual children to come and place more shapes into the correct sets.
- Place two sorting hoops on the floor. Label them 'squares' and 'not squares'. Ask a child to take a shape from the 'feely' bag. Ask: *Is this shape a square?* Ask the child to place the shape into the correct hoop.
- Repeat several times.
- Teach children the action rhyme *Draw a square* (24).

Explore in groups

Maths Foundation Reading Anthology
(adult-led)

- Together, look at pages 16 and 17 (2D shapes) of the *Reading Anthology*. Discuss the examples of squares in the picture. Ask children to draw some things that are square-shaped. They can use ideas from the *Anthology*, or their own ideas.

Pairs

- Provide shape cards from PCMs 18–21. Encourage children to play Pairs (see Generic games rules on page 220). Each pair or group will need any six square cards, two circle cards, two triangle cards and two rectangle cards. Children match pairs of cards according to shape.

Squares and not squares

- Provide a set of circle, triangle, square, and rectangle cards from PCMs 18–21. Invite children to sort the cards into two sets: 'squares' and 'not squares'.

Variation: use 2D geometric shapes instead of the cards.

Find the squares

- Provide copies of PCM 22. Encourage children to colour all the squares green. When finished, you could ask children to draw three different squares on the back of their sheet.

Session 5: Recognise and name rectangles

Children begin to recognise and name rectangles, including in different sizes and orientations.

You will need:

Shape set Digital Tool; 2D geometric shapes; non-transparent ('feely') bag; two sorting hoops; labels: 'rectangles' and 'not rectangles'; *Maths Foundation Reading Anthology*; paper; coloured pencils or crayons; PCMs 18–21; interlocking cubes; *Maths Foundation Activity Book A*

Getting started

- Before the session, place some 2D geometric shapes in the 'feely' bag (include more rectangles than any other shape).
- Show children a rectangle. Briefly discuss its properties: a 2D, flat shape with four straight sides, but not all sides the same size. At this stage, do not expect children to recall these properties. They should just be able to recognise and name rectangles, and

distinguish between shapes that are rectangles and shapes that are *not* rectangles, especially squares.
- Rotate the rectangle to show it in different orientations.
- Show other rectangles of different sizes. Constantly reinforce the words 'rectangle', '2D' and 'flat' so that children begin to recognise and classify rectangles as 2D, flat shapes (including rectangles in different orientations.)

Teaching

- Show children a shape that is *not* a rectangle, such as a triangle. Ask: *Is this shape a rectangle? What's the same about this shape and a rectangle? What's different about them?*
- Repeat for other 2D shapes that are *not* rectangles.
- Display the Shape set Digital Tool. Add two sets: one set labelled 'rectangles', the other set labelled 'not rectangles'.
- Place different 2D shapes into the correct sets. Click on individual shapes, especially rectangles, to change their colour, size and orientation.
- Ask individual children to come and place more shapes into the correct sets.
- Place two sorting hoops on the floor. Label them 'rectangles' and 'not rectangles'. Ask a child to take a shape from the 'feely' bag. Ask: *Is this a rectangle?* Ask the child to place the shape into the correct hoop.
- Repeat several times.
- Teach children the action rhyme *Draw a rectangle* (25).

Explore in groups

Maths Foundation Reading Anthology
(adult-led)

- Together, look at pages 16 and 17 (2D shapes) of the *Reading Anthology*. Discuss the examples of rectangles in the picture. Ask children to draw some things that are rectangle-shaped. They can use ideas from the *Anthology*, or their own ideas.

Rectangles and not rectangles

- Provide circle, triangle, square and rectangle cards from PCMs 18–21. Encourage children to sort the cards into two sets: 'rectangles' and 'not rectangles'.

Teaching notes

Variation: use 2D geometric shapes instead of the cards.

Making rectangles

- Provide interlocking cubes. Invite children to use them to make rectangles (rectangular frames). Ask: *How did you decide how many cubes to use for each side?* Do they recognise that each pair of opposite sides must be the same length? Do they recognise that if they use the same number of cubes for each side, it will make a square?

Maths Foundation Activity Book A (adult-led)

Page 21 – Squares and rectangles

Session 6: Compare 2D shapes

Children compare two 2D shapes, saying what is the same and what is different.

You will need:

Maths Foundation Reading Anthology; Shape set Digital Tool; 2D geometric shapes; non-transparent ('feely') bags; PCMs 18–21

Getting started

- Begin the session by saying some of the shape action rhymes: *Draw a circle* (22), *Draw a triangle* (23), *Draw a square* (24) and/or *Draw a rectangle* (25).

Teaching

- Show pages 16 and 17 (2D shapes) of the *Reading Anthology*. Ask individual children to come and point to examples of triangles.
- Then point to each of the triangles. Ask: *What's the same about all of these triangles? What's different?* Reinforce the word 'triangle'.
- Repeat for circles, squares and rectangles. For the pictures of real-world examples of 2D shapes, trace round the object so that children understand you are referring to the 2D shape.
- Display the Shape set Digital Tool. Place a circle and a triangle on the screen. Point to each shape. Ask: *What is this shape? And what is this shape called?*
- Ask: *What is the same about them?* Discuss suggestions, e.g. they are the same colour,

they are about the same size, they are both 2D, flat shapes.
- Ask: *What is different?* Discuss suggestions, e.g. a circle is round, a circle has one side; a triangle has three sides, all the sides of a triangle are straight.
- Click 'Clear all', repeat to compare other pairs of shapes: circle and square; circle and rectangle; triangle and square; triangle and rectangle; square and rectangle.

Explore in groups

Collect the four shapes

- Provide 'feely' bags, each containing 12 2D geometric shapes: 2 circles, 2 triangles, 2 squares, 2 rectangles and 4 other shapes. Invite children to take a bag and play a game in pairs. They take turns to feel inside for a shape. Each child is aiming to collect a circle, triangle, square and rectangle, without removing any other shapes from the bag.

Collect all the shapes

- Shuffle a set of 5 circle cards from PCM 18 and 5 triangle cards from PCM 19. Spread them out face down. Encourage children to play a game in pairs. One child is aiming to collect all the circle cards, the other child is aiming to collect all the triangle cards. They take turns to turn over a card. If the card shows their shape, they keep it. If not, they put it back. The winner is the first child to collect the 5 cards.

Variations:

- Use 5 circle cards from PCM 18 and 5 square cards from PCM 20.
- Use 5 circle cards from PCM 18 and 5 rectangle cards from PCM 21.
- Use 5 triangle cards from PCM 19 and 5 square cards from PCM 20.
- Use 5 triangle cards from PCM 19 and 5 rectangle cards from PCM 21.
- Use 5 square cards from PCM 20 and 5 rectangle cards from PCM 21.

Shape pictures

- Provide lots of 2D geometric shapes (circles, triangles, squares and rectangles). Invite children to make pictures using shapes. Their picture could be linked to the theme 'plants', or anything they like. Ask questions such as: *What is this shape called? How do you know? How is it different from this shape here?* You could take photos of children's completed pictures.

Session 7: Sort 2D shapes

Children sort different 2D shapes into groups.

You will need:

2D geometric shapes; four sorting hoops; labels: 'circles', 'triangles', 'squares', 'rectangles'; non-transparent ('feely') bags; large books; PCMs 18–21; sand tray; 4 dishes; *Maths Foundation Activity Book A*; coloured pencils or crayons

Getting started

- Before the session, place a selection of 2D geometric shapes (circles, triangles, squares and rectangles) in a 'feely' bag. Keep one of each shape to use at the start of *Teaching*.
- Say some of the shape action rhymes: *Draw a circle* (22)*, Draw a triangle* (23)*, Draw a square* (24) and/or *Draw a rectangle* (25).

Teaching

- Hold up a circle. Ask: *What is this shape called? How do you know it's a circle?* Briefly discuss the properties: a 2D, flat, round shape with one curved side.
- Place a sorting hoop on the floor. Label it: 'circles'. Place the circle in the hoop.
- Repeat for triangle, square and rectangle.
- Show children the 'feely' bag containing circles, triangles, squares and rectangles.
- Choose a child. Say: *Without looking, feel for a shape inside the bag and tell me its name.* The child does so, then holds up the shape. Ask: *Is Rosa right?* Ask the child to place the shape into the correct hoop.
- Repeat with different children.
- Count how many of each there are of each shape. Briefly discuss the similarities and differences between the shapes.

Explore in groups

Name the shape

- Before the activity, prepare some 'feely' bags. Put a selection of 2D geometric shapes in each bag (circles, triangles, squares and rectangles). Provide large books that can stand upright to create a screen. Encourage children to play a game in groups. They take turns to secretly take a shape out of their group's bag. They place it behind a book and slowly reveal the shape. The other children have to name the shape as soon as they can.

Sort the shapes

- Provide shuffled selections of circle, triangle, square and rectangle cards from PCMs 18–21. Invite children to sort the cards. How many different ways can they sort the cards into two, three or four groups?

Sand station

- Before the activity, bury some 2D geometric shapes in the sand. Include circles, triangles (different types), squares, rectangles, and a few other shapes. Provide four dishes. Let children hunt for the shapes. When they have found some, encourage them to start sorting them into groups using the dishes. Ask them to tell you about the shapes they have found, and how they have chosen to sort them.

Maths Foundation Activity Book A (adult-led)

Page 22 – Sort

Session 8: Make patterns using 2D shapes

Children observe, copy, continue and create 'AB' repeating patterns using 2D shapes.

You will need:

Shape set Digital Tool; PCM 23 (possibly enlarged to A3); paper; resources for making 2D shape patterns (circles, triangles, squares and rectangles), e.g. shape stickers, self-inking shape stamps, 3D geometric plastic cylinder, pyramid, cube, cuboid, paints or stamp pads; 2D geometric shapes; *Maths Foundation Activity Book A*; coloured pencils or crayons

Teaching notes

Getting started

- Say some of the shape action rhymes: *Draw a circle* (22), *Draw a triangle* (23), *Draw a square* (24) and/or *Draw a rectangle* (25).

Teaching

- Display the Shape set Digital Tool. Place six circles in a row: red, blue, red, blue, red, blue.
- Say: *I have made a pattern of six circles.* Point: *Red, blue, red, blue, red, blue. I want to carry on with this pattern. What colour circle should I have next?* Emphasise the words 'pattern' and 'next'.
- Discuss responses. Place another red circle in the sequence. *What goes next in my pattern?* Continue until there are 10 circles. Say: *In this pattern I have only used circles. In this pattern it's the colour that changes: red, blue, red, blue, red, blue … This is called a repeating pattern because the colours repeat more than one time.* Emphasise the words 'repeating pattern'.
- Repeat above, using circles, triangles, squares or rectangles to make another AB pattern based on colour. Show three full units of repeat (AB AB AB), before asking children to continue the pattern.
- Now use two different shapes to create an AB pattern based on shape, e.g. circle, triangle, circle, triangle, circle, triangle. Make sure that all the circles are the same colour and size, and all the triangles are the same colour and size. Ask: *What do you notice about this pattern?* Discuss responses. *What will come next in my pattern?* Place the next shape. Continue until there are 10 shapes in the pattern (5 units of repeat).
- Repeat several times, making other AB patterns using pairs of shapes.

Explore in groups

Shape patterns (A)

- Provide sheets of paper and suitable resources for making 2D shape patterns (circles, triangles, squares and rectangles),

e.g. shape stickers, self-inking shape stamps; 3D geometric plastic cylinder, pyramid, cube, cuboid; paints or stamp pads. (If using 3D geometric plastic shapes, children use an end face of a cylinder as a circle 'stamp', the triangular face of a pyramid as a triangle 'stamp', any face of a cube as a square 'stamp', and a rectangular face of cuboid as a rectangle 'stamp'.) Invite children to use the resources to create AB shape patterns.

Shape patterns (B)

- Provide 2D geometric shapes. Invite children to use the shapes to create AB shape patterns.

Continue my pattern (adult-led)

- Provide a pile of pattern tracks from PCM 23. Encourage children to play in pairs. Each child takes a pattern track and makes an AB 2D shape pattern in the first six boxes of their track (e.g. square, triangle, square, triangle, square, triangle). Children then swap tracks and complete their partner's pattern. When complete, encourage children to turn each pattern 90°, then 180° and then 270°. Discuss the pattern with the children: *What do you notice? What is the same? What is different?*

Maths Foundation Activity Book A (adult-led)

Page 23 – Shape patterns

Assessment opportunities

Assess children's learning against the objectives for this unit, using the guidance on formative assessment on pages 24–25, and record your observations in the Unit 5 progress tracking grid on page 28. The relevant pages of *Activity Book A* can also be used for assessment.

Can the children:

- recognise and name circles, triangles, squares and rectangles in different sizes and orientations?
- compare two 2D shapes?
- sort 2D shapes into groups?
- use 2D shapes to observe, copy, continue and create AB repeating patterns?

Unit 6 Numbers to 10 (A)

Theme: Food

Overview

In Units 1 and 2, children learned about numbers 1 to 5. In this unit, the number range is extended to 10. Children count from 1 to 10 forwards and backwards. They count sets of objects, and things that cannot be counted. They compare quantities and numbers up to 10. They start to learn to read and write the numerals 6 to 10.

Children are also encouraged to look at a small number of objects and know how many there are without counting. This is called *subitising*. It is a key skill for counting accurately and confidently (and, later, for adding and subtracting).

Learning objectives

Number – Counting and understanding numbers	• *1a Say and use the number names in order in familiar contexts such as number rhymes, songs, stories, counting games and activities, from 1 to 5,* **then 1 to 10.** • *1b Recite the number names in order, continuing the count forwards or backwards, from 1 to 5,* **then 1 to 10.** • *1c Count objects from 1 to 5,* **then 1 to 10.** • *1d Count in other contexts such as sounds or actions from 1 to 5,* **then 1 to 10.**
– Reading and writing numbers	• *2a Recognise numerals from 1 to 5,* **then 1 to 10.** • *2b Begin to record numbers, initially by making marks, progressing to writing numerals from 1 to 5,* **then 1 to 10.**
– Comparing and ordering numbers	• *3a Use language such as more, less or fewer to compare two numbers or quantities from 1 to 5,* **then 1 to 10.**

Learning objectives in italics have been taught previously. In this unit they are consolidated and/or extended.

Learning objectives (or parts of objectives) in bold are taught for the first time in this unit.

Vocabulary

number, count, count on, count forwards, count back, count backwards, one, two, three, …, eight, nine, ten, next, after, before, how many, compare, more, less, fewer, the same, more than, less than

Making connections

English: Food

Science: What plants and animals need

Preparation

You will need:

- *Maths Foundation Reading Anthology*, pages 10–13.
- *Maths Foundation Activity Book B*, pages 2–5.
- Unit 1 Session 1 slides.
- Unit 6 slides.
- Digital Tool: Counting.
- PCMs 1–13.
- Number fluency games and activities: *Finger counting* (1), *Show me* (2) (page 198).
- Rhymes and songs: *Here is the beehive* (3), *This old man* (5), *Zoom, zoom, we're going to the moon* (8), *Ten little fishies* (16) (pages 209–214).
- Large 1–10 number bunting and/or large number track.
- Puppet (any character or animal).
- Beads and laces.
- Interlocking cubes.
- 1 to 6 dot dice.
- Small counters.
- Percussion instruments, e.g. chime bars, drums and sticks.
- PE equipment, e.g. balls, net (or hoop), skittles, individual skipping ropes, beanbags, bucket, quoit set.
- Various classroom objects and displays that have numbers, e.g. a clock, number bunting, room number, wall displays.

- Painting and drawing station: paper, pencils, coloured pencils and/or crayons, paint, water, paintbrushes, aprons, cards 6–10 from PCM 1.
- Counting collections station: a selection of different small-world resources (e.g. people, animals and sealife, transport) and counting apparatus (e.g. cubes, counters), objects associated with food to use as counting apparatus, small pots or cups, cards 6–10 from PCMs 2 and 5.
- Sand station: sand tray/pit, small pots, cards from PCM 3, numeral cards 6–10 from PCM 5.

Before starting

Display number bunting or a large number track showing numerals 1 to 10. Each numeral should have a matching picture (e.g. 9 dots below the numeral 9). Refer to the bunting/track regularly throughout the unit.

Where appropriate, use objects linked to the theme of 'food' as counting apparatus.

Before teaching this unit, look back at the Assessment section at the end of Unit 2 (page 50), to identify the prerequisite learning for this unit. If children are not yet confident with numbers 1 to 5, do not extend to numbers up to 10.

Maths background

The word 'fewer' is used when talking about people or things in the plural. The word 'less' is used when talking about things that are uncountable, or have no plural. It is important to use the correct mathematical vocabulary throughout this unit. When comparing two sets of objects, use the words 'more' and 'fewer'. When comparing two numbers, use 'more' and 'less'.

Session 1: Count forwards and backwards from 1 to 10

Children count forwards from 1 to 10, and say which number comes next and before. They understand that the numbers are said in a certain order: *stable order*. They also count backwards from 10 to 1.

You will need:

Unit 1 Session 1 slides; puppet; paper; coloured pencils or crayons, or paint, water, paintbrushes, aprons

Getting started

- Say the rhyme *Here is the beehive* (3), to practise counting forwards from 1 to 5. Display the Unit 1 Session 1 slides. Click through the 6 slides as you all say the rhyme together.
- Say the action rhyme *Zoom, zoom, we're going to the moon!* (8) to practise counting backwards from 5 to 1.

Teaching

- Say the numbers 1 to 10 in order. Then ask the children to count on from 1 to 10 with you. Repeat several times. Say: *We have counted on, or forwards, from 1 to 10.* Emphasise the words 'on' and 'forwards'.
- Show the puppet. Ask children to help the puppet to count on from 1 to 10. Repeat several times.
- Emphasise the words 'next' and 'after': *The next number after 6 is 7.*
- Ask: *Who can tell the puppet the number that comes after 5 when counting on?* Repeat for other numbers.
- Do the same for the word 'before': *The number that comes before 7 is 6. Who can tell the puppet the number that comes before 8?* Repeat for other numbers.
- Next, slowly count backwards from 10 to 1. Repeat. Ask: *What do you notice about the way I counted this time? What was the same? What was different?*
- Discuss: *I used the same numbers to count. But instead of counting forwards from 1 to 10, I counted backwards from 10 to 1.* Emphasise the words 'forwards' and 'backwards'. Count backwards from 10 to 1 again.
- Ask children to help the puppet to count backwards from 10 to 1. Repeat several times.
- Say the action rhyme *Ten little fishies* (16). Repeat several times. Then ask the children to join in.

Explore in groups

Repeat Teaching (adult-led)

- Repeat the *Teaching* activity with smaller groups of children. Encourage each child to join in with counting on from 1 to 10. Say: *The puppet can't remember what number comes before 2 [or any number 3 to 10]. Can*

you help? Repeat for the number that comes *after* 1 (or any number 2 to 9). Also practise counting back from 10 to 1, but do not ask the children to say the number *before* or *after*.

Puppet mistakes (adult-led)

- Explain that the puppet is going to count forwards from 1 to 10. Make the puppet say the numbers 1 to 10, but make a deliberate mistake, e.g. *1, 2, 3, 4, 5, 6, 8, 7, 9, 10.* Ask the children to spot the mistake. Repeat several times. Each time, leave out a number, or say the same number twice, or say the numbers out of order. Repeat the activity, this time counting backwards from 10 to 1.

Painting and drawing station

- Invite children to draw or paint a picture based on the rhyme *Ten little fishies*. Ask questions focused on the number of fish, e.g. *How many fish have you drawn? Can you remember how many are in the rhyme? Would you like to draw all ten fish?*

Session 2: Count up to 10 objects

Children count groups of up to 10 objects, giving one number name to each object: *one-to-one correspondence*.

They further develop their understanding that the last number spoken matches the quantity for that set: *cardinality*. They also understand that it doesn't matter in which order we count a group of objects: *order-irrelevance*.

You will need:

Unit 6 Session 2 slides; PCM 1; cards 6–10 from PCMs 3 and 4; interlocking cubes; beads and laces; paper; coloured pencils or crayons, or paint, water, paintbrushes, aprons; *Maths Foundation Reading Anthology*

Getting started

- Revise counting up to 5 objects. Display slide 1. Point to a plate. Ask children to tell you how many pieces of food are on the plate. Take responses. Reinforce the cardinal value by pointing to each piece of food in turn, saying, e.g. *1, 2. There are 2 pieces of chicken.* Repeat for all five plates.

Teaching notes

Teaching

- Display slide 2. Introduce the number name 'six', and how this is represented by 6 objects (cakes).
- Point to each cake in turn, saying one number name for each cake: *1, 2, 3, 4, 5, 6. There are 6 cakes.*
- Count the cakes again but in a different order, e.g. anticlockwise. Remind children that the order in which the cakes are counted doesn't matter: *order irrelevance.*
- Repeat for slides 3 to 6.
- Display slide 7. Ask individual children to come and point to the basket that shows 6/7/8/9/10 vegetables.
- Give each child a card (1–10) from PCM 1. Say: *I will say a number. If the number of things on your card matches the number, stand up and show me your card.* Say a number, e.g. 8. Repeat until you have said all the numbers from 1 to 10.

Explore in groups

Interlocking cube towers

- Place multiple copies of cards 6–10 from PCM 3 (cubes) next to a collection of loose interlocking cubes. Prompt children to make cube towers to match the cubes on the cards. Encourage children to check each other's towers by counting.

Bead strings

- Place multiple copies of cards 6–10 from PCM 4 (ten-frames) next to a collection of beads and laces. Prompt children to make bead strings to match the number of dots on the cards. Encourage children to check each other's bead strings by counting.

Painting and drawing station

- Provide cards 6–10 from PCM 1 (objects). Invite children to choose a card, and draw or paint a picture of that number of items. Encourage them to draw/paint items linked to the theme 'food', e.g. pieces of fruit.

Maths Foundation Reading Anthology (adult-led)

Pages 10 and 11 – How many animals?

- Together, look at pages 10 and 11 (How many animals?) of the *Reading Anthology*. Discuss the pictures with the children. Point to, and count, groups of up to 10 animals.

Session 3: Count up to 10 children and objects

Children count groups of up to 10 objects. They continue to develop their understanding that the number of objects in a group does not change if the objects are rearranged: *conservation.*

You will need:

cards 6–10 from PCMs 1 and 2; selection of different small-world resources and counting apparatus; *Maths Foundation Reading Anthology*; *Maths Foundation Activity Book B*; pencils

Getting started

- Use the Number fluency activities *Finger counting* (1) and *Show me* (2), to practise counting forwards and backwards from 1 to 10, and counting groups of up to 10 objects.

Teaching

- Remind children that they have been learning number names and counting groups of up to 10 objects.
- Choose a child to stand up. Ask all the children: *How many children are standing?* Choose a second child to stand next to the first. Ask the same question. Repeat, until there are 4 children standing in a row.
- Swap the positions of the children. Ask: *How many children are standing?*
- Now position the children so that they are not in a row, e.g. each child stands in a different area of the room. Again, ask: *How many children are standing?*
- Repeat several times, using up to 10 children.
- Show cards 6–10 from PCMs 1 and 2. Ask children to count the objects on each card. Repeat for groups of up to 10 different small-world resources and counting apparatus.

Explore in groups

Pairs

- Provide sets of cards 6–10 from PCMs 1 (objects) and 2 (fingers). Encourage children to play Pairs (see Generic games rules on page 220). Each pair or group will need one set of object cards and one set of finger cards.

Counting collections station

- Place cards 6–10 from PCM 2 (fingers) next to a variety of small-world resources. Prompt children to make collections of small-world resources to match the fingers on the cards. Encourage children to check each other's groups by counting.

Maths Foundation Reading Anthology (adult-led)

- Together, look at pages 12 and 13 (The fruit stall) of the *Reading Anthology*. Discuss the picture with the children. Point to, and count, groups of up to 10 fruit. Encourage the children to find and count other groups of up to 10 fruit.

Maths Foundation Activity Book B (adult-led)

Page 2 – Count and match

Session 4: Recognise up to 6 objects: subitising

Children begin to look at a small number of objects and know how many there are without counting: *subitise.*

You will need:

Unit 6 Session 2 slides; PCMs 9 and 10; cards 1–6 from PCMs 1–4; 1 to 6 dot dice; counting apparatus; small pot or cup

Getting started

- Display slide 1. Ask children to tell you how many pieces of food there are on each plate.

Teaching

- Explain that sometimes we can just look at a group of objects and know how many there are without having to count.

- Shuffle the subitising cards from PCM 9 (dots in regular patterns). Briefly hold up a card. Ask children to say how many dots there are. Show the card again and count the dots.
- Repeat for the remaining five cards. Shuffle the cards and repeat.
- Say: *I'm going to do that again, but I will use some cards that you've seen before.* Shuffle cards 1–6 from PCMs 1–4. Repeat as above, several times.
- Say: *It's not always easy to see how many objects there are.* Repeat above, briefly holding up each of the subitising cards from PCM 10 (dots in irregular formations).

Explore in groups

Match the dice

- Provide a 1 to 6 dot dice and the cards from PCM 9. Spread the cards face up on the table. Encourage children to roll the dice, and turn over the card with the matching number of dots. They continue until all cards are turned over.

Variations: use the cards from PCM 10 (dots in irregular formations). Alternatively, turn the activity into a paired or group game. Each child has a set of cards from PCM 9 or 10. Children take turns to roll the dice. The first child to turn over all 6 of their cards is the winner.

Pairs

- Provide cards from PCMs 9 and 10. Encourage children to play Pairs (see Generic games rules on page 220). Each pair or group will need all 12 cards from PCMs 9 and 10.

Counting collections station (adult-led)

- Secretly take up to 6 counting apparatus, e.g. 5 cubes. Hide them in your hands (or in a small pot or cup). Briefly show the children what's in your hands. Ask: *How many cubes am I holding?* Allow each child to tell you how many they think there are. Place the cubes on the table and count them together. Repeat several times. Use a range of different counting apparatus.

Teaching notes

Session 5: Count sounds and actions to 10

Children continue to develop their understanding that they can count non-physical things such as sounds, actions and remembered or imaginary objects: *abstraction*.

You will need:

percussion instruments; PCM 3; PE equipment; sand tray; small pots or cups

Getting started

- Sing the action song *This old man* (5) (all 10 verses).

Teaching

- Remind children that numbers can be used for counting sounds and actions as well as objects.
- Assign each child an action, e.g. jump in the air, tap their head, tap their shoulders, clap their hands, touch their knees, star jump, frog jump, hop.
- Say: *Each time I make a sound, do your action once.*
- Hit a drum (or other percussion instrument) 3 times. Allow enough time after each hit for children to perform their action once. Reinforce: *I hit the drum 3 times and you did your action 3 times.*
- Repeat, hitting the drum twice, then 5 times, then 4 times.
- Then ask children to listen very carefully. Say: *I will hit the drum a number of times. When I've finished, I'll point to you. You then do your action that number of times.*
- Hit the drum 5 times, then point to the children. Once all the children have performed their action 5 times say: *Well done! I hit the drum 5 times, and you did your action 5 times.*
- Repeat several times, hitting the drum different numbers of times from 1 to 10.

Explore in groups

Music makers

- Provide a selection of percussion instruments. Spread the cards from PCM 3 (cubes) face down on the table. Invite children to turn over a card and count the cubes. They then make that number of sounds using an instrument.

Outdoor play (adult-led)

- Create a simple obstacle course. Each 'obstacle' must be completed a given number of times, e.g. kick a ball into a net (or hoop) 6 times; knock down 7 skittles; skip 8 times using a skipping rope; throw or drop 9 beanbags into a bucket; throw or drop 10 quoits over the peg.

Sand station

- Shuffle a set of cards from PCM 3 (cubes). Place them face down in a pile near the sand tray. Also place small pots near the tray. Encourage children to take the top card from the pile and count the cubes. They then fill that number of pots with sand.

Session 6: Match a numeral to a set of 1 to 10 objects

Children continue to use the numerals 1 to 5 to represent quantities. They begin to use the numerals 6 to 10.

You will need:

Unit 6 Session 6 slide; Counting Digital Tool; cards 6–10 from PCMs 1–4; PCMs 5 and 11; selection of different small-world resources and counting apparatus; small pots or cups; container of small counters; *Maths Foundation Activity Book B*; pencils

Getting started

- Display slide 1. Point to each number in turn, saying the number name. Ask children to repeat the number after you. Discuss each number, highlighting its 'shape'. If appropriate, discuss the context in which the number is used.
- Ask children to talk about their experiences of numbers: *Who can tell me some other numbers they know? Where have you seen numbers? What are they for? How are they used?*

Teaching

- Display the Counting Digital Tool. Set it to show the beach scene with the crab and fish.

- Hold up numeral card 6 from PCM 5. Say: *I am going to place some fish in the scene. When there are 6 fish, hold up 6 fingers and shout: '6!'*
- Slowly place 6 fish into the scene. Ask: *Are there 6 fish? Let's count them to check.* Count the fish together, pointing to each one. Click on the fish card to the right of the scene to display the numeral 6. Compare the numeral 6 on the screen with the 6 on the numeral card from PCM 5.
- Stand with your back to the children. Hold up the numeral card so they can see it. Slowly trace over the '6' with your finger. Say: *This is how we write the number 6. Now you write 6 in the air. Show me that again.*
- Click 'Clear all'. Click the fish card to hide the numeral 0.
- Repeat for numbers 7 to 10.

Explore in groups
Counting collections station

- Spread numeral cards 6–10 from PCM 5 face up on the table. (You could also include some of the cards 1–5.) Next to each card place small pots. Suggest that children place small-world resources or counting apparatus into each pot to match the number on the card.

Ten-frames

- Spread numeral cards 6–10 from PCM 5 face up on the table. (You could also include some of the cards 1–5.) Provide a pile of blank ten-frame cards from PCM 11 and a container of small counters. Invite children to take a numeral card and represent the number by placing counters on a ten-frame.

Pairs

- Provide sets of cards 6–10 from PCMs 1 (objects) and 5 (numerals). Encourage children to play Pairs (see Generic games rules on page 220). Each pair or group will need one set of objects cards and one set of numeral cards.

Variation: use cards 6–10 from PCMs 2, 3 or 4 instead of PCM 1.

Maths Foundation Activity Book B (adult-led)

Page 3 – Count and match

Session 7: Write the numerals 6 to 10
Children begin to write the numerals 6 to 10.

You will need:

numeral cards 6–10 from PCM 5; various classroom objects and displays that have numbers; Unit 6 Session 7 slides; PCMs 7, 8, 12 and 13; pencils; coloured pencils; sand tray; paper; paint, water, aprons; *Maths Foundation Activity Book B*

Getting started

- Hold up numeral card 7 from PCM 5. Ask children to say the number.
- Ask children to point to other examples of the number 7 in the classroom and/or outdoors.
- Repeat for the other numeral cards 6–10.

Teaching

- Display slide 1. Say: *We are going to learn how to write the number 6.*
- Use the row of numerals to demonstrate how to write the number 6. Slowly tracing over each numeral and explain how it is written. Write two extra number 6s at the end of the row.
- You could use the following rhyme: *Draw a curve, then make a loop. There are no tricks to write a six.*
- Ask children to write an imaginary 6 in the air with you, as you describe the formation of the numeral.
- Repeat for the numerals 7, 8, 9 and 10 using slides 2–5 in turn. You could use the following rhymes.
  - 7: *Across the sky, then down like a feather. That's how to write a seven.*
  - 8: *Write an S, and do not wait. Go back up, and that's an eight.*
  - 9: *Make a loop, and then a line. That's how to write a nine.*
  - 10: *Straight line down, that makes a one. Then around and round you run. That makes a ten, oh what fun.*

Explore in groups
Practising numeral formation

- Give each child a pencil and a copy of PCM 12. Encourage them to practise writing the numerals 6 to 10.

Teaching notes

Variation: use PCM 7 to practise writing the numerals 1 to 5.

Practising numeral formation and recognising cardinal values

- Before the activity, prepare copies of PCM 13. On each sheet, circle two numbers. Use a different coloured pencil for each circle, e.g. a blue circle around 6, and a red circle around 8. Give each child one of these sheets and two coloured pencils to match the colours of the circles. Ask children to trace each circled number in the corresponding colour. Then invite them to colour that number of stars in the same colour (e.g. they colour 6 stars blue and 8 stars red).

Variation: use PCM 8 instead. Children practise writing the numbers 1 to 5 and colour balloons to match.

Sand station

- Place a set of numeral cards 6–10 from PCM 5 near to the sand tray. Slightly dampen the sand and smooth it over. Let children write each of the numerals 6 to 10 in the sand, using the numeral cards as a guide.

Painting and drawing station

- Encourage children to finger-paint the numbers 6 to 10. Place a set of numeral cards 6–10 from PCM 5 on the table for children to refer to.

Maths Foundation Activity Book B (adult-led)

Page 4 – Trace and write

Session 8: Compare two quantities and numbers to 10

Children continue to use the words 'more' or 'fewer' to compare two quantities, and the words 'more' or 'less' to compare two numbers.

You will need:

cards 6–10 from PCMs 1 and 5; Unit 6 Session 8 slides; PCM 6; coloured pencils or crayons; 6 counters (or similar); stapler or glue; *Maths Foundation Activity Book B*

Getting started

- Hold up two of the cards (6–10 only) from PCM 1. Ask, for example: *Are there more bikes or cars?*
- Repeat several times. Hold up two cards each time. Ask questions such as: *Are there fewer … or …? Are there more … or …? Which card shows more? Which card shows fewer?*

Teaching

- Display slide 1. Ask: *How many red counters are there? How many yellow counters? Which is more: 6 or 3?*
- Point to the numbers 3 and 6 on the number track: *6 is more than 3.* Emphasise the word 'more'.
- Repeat for slides 2 to 5. Each time, use the number track to show the position of each number in the counting sequence: *… is more than …*
- Display slide 1 again. Ask: *How many red counters? How many yellow counters? There are fewer yellow counters than red counters.* Emphasise the word 'fewer'.
- Point to numbers 3 and 6 on the number track: *3 is less than 6.* Emphasise the word 'less'.
- Repeat for slides 2 to 5. For each slide, ask children to say which colour counters there are 'fewer' of. Use the number track to show the position of each number in the counting sequence: *… is less than …*
- Hold up two of the numeral cards (6–10 only) from PCM 5. Ask, for example: *Which is more: 6 or 9?*
- Repeat several times. Hold up two cards each time. Ask questions such as: *Which is less: … or …? Which is more: … or …?*

Explore in groups

Which tree has more?

- Before the activity, prepare copies of PCM 6. On each sheet, write a number from 6 to 10 above each tree. Invite children to draw the corresponding number of apples (or other fruit) on each tree. Then prompt them to draw a ribbon around the tree with *more* (or *fewer*) apples/fruit.

Variation: write a number above the first tree only. Children draw that number of apples on the tree.

Then ask the child to draw more (or fewer) apples on the second tree and write the corresponding number above the tree.

Comparing numbers (adult-led)

- Shuffle two sets of numeral cards 6–10 from PCM 5. Spread the 10 cards face down on the table. Ask two children to each turn over a card. Ask: *Who has the number that is more?* Give the child a counter (or similar). If two children turn over matching cards, and say that they are 'the same', give each child a counter. Continue until all 10 cards have been used.

Variation: ask: *Who has the card with the number that is less?* Give the child a counter.

More or less cards

- Before the activity, prepare some double-sided cards. Use numeral cards 6–10 from PCM 5. Staple or glue pairs of cards back-to-back. You will need at least 3 or 4 double-sided cards for each child. Invite children to look at the number on each side of the card and circle the number that is *more*.

Variation: ask children to circle the number that is *less*.

Maths Foundation Activity Book B (adult-led)

Page 5 – Trace and draw

When children have completed the page, point to two egg cartons, e.g. 6 and 8. Ask them to point to the carton with *more* (or *fewer*) eggs. Repeat, pointing to two numbers and asking the child to say which number is *more* (or *less*).

Assessment opportunities

Assess children's learning against the objectives for this unit, using the guidance on formative assessment on pages 24–25, and record your observations in the Unit 6 progress tracking grid on page 29. The relevant pages of *Activity Book B* can also be used for assessment.

Can the children:

- show an understanding of the five key counting principles and of conservation of number?
- recite the number names 1 to 10 in order, counting forwards and backwards?
- say which number comes next and which number comes before, when counting forwards?
- count up to 10 objects and things that cannot be touched?
- visually recognise a quantity of 6 or fewer: *subitising*?
- recognise and attempt to write (as numerals) numbers 1 to 10?
- use the words 'more' or 'fewer' to compare two quantities from 1 to 10?
- use the words 'more' or 'less' to compare two numbers from 1 to 10?

Unit 7 Addition as combining two sets

Theme: My home

Overview

There are two different ways of looking at addition:

1) combining two quantities into a single quantity to work out the total – putting together by counting all

2) a given quantity is increased by another quantity – becoming greater by counting on.

This unit focuses on the first structure of addition: counting all. The key language in this unit includes: 'How many altogether?', 'How much altogether?' and 'What is the total?' This unit also introduces children to the terms 'add', 'plus' and 'is equal to'.

Learning objectives

Number – Understanding addition and subtraction	• **4a In practical activities and discussions begin to use the vocabulary involved in addition: combining two sets [and counting on].**

Learning objectives (or parts of objectives) in bold are taught for the first time in this unit.

[This part of the learning objective is not taught in this unit.]

Vocabulary

number, count, one, two, three, …, eight, nine, ten, how many, how much, altogether, total, add, plus, makes, is equal to

Making connections

English: My house

Science: Habitats

Preparation

You will need:

- Maths *Foundation Reading Anthology*, pages 2, 3, 8, 9 and 10.
- *Maths Foundation Activity Book B*, pages 6–9.
- Unit 7 slides.
- Digital Tools: Tree, Counting.
- PCMs 1–3, 5–6, 9–11, 14 and 15.
- Number fluency games and activities: *Finger counting* (1), *Show me* (2), *Match the cards* (14) (pages 198–201).
- Rhymes and songs: *Ten little fishies* (16) (page 214).

- Large 1–10 number bunting and/or large number track.
- Chalk.
- Interlocking cubes.
- Counters (or similar).
- Double-sided (two-colour) counters (or dried lima/butter beans with one side painted).
- Trays, small pots or cups, plates or dishes.
- Plastic or real fruit: lemon, apples, oranges, strawberries and bananas.
- 2 'fruit' boxes.
- Basket.

- Glue.
- 2 large 1 to 6 dot dice (with each 6-dot face covered with paper and the 3-dot pattern drawn on).
- Non-transparent ('feely') bag.
- Washing line (or string).
- 10 clothes pegs.
- 10 socks (or similar) in two different colours (5 of each).
- Counting collections station: a selection of different small-world resources (e.g. people, animals and sealife, transport), small-world playmats and 'environments', number track from PCM 15.
- Painting and drawing station: paper, pencils, coloured pencils or crayons, paint, water, paintbrushes, aprons.

Before starting

Display number bunting or a large number track showing numerals 1 to 10. Each numeral should have a matching picture (e.g. 3 dots below the number 3). Refer to it regularly throughout the unit.

Children will get the most out of this unit if they already have experience of:

- counting groups of up to 10 objects
- recording the number of objects in a group by making marks, drawing pictures or by writing numerals.

Common difficulties

If some children still need more practice counting a group of objects, find lots of opportunities to ask them to count. For example, ask them to check that you have put the correct number of snacks on a table. Encourage them to pick up and move objects when counting, e.g. blocks from one cup to another.

Maths background

Mathematically, the word 'sum' means the same as 'total': it is the result of adding two or more numbers. 'Sum' is often used informally to refer to any number sentence: *calculation*. Strictly speaking it should apply only to addition.

Session 1: Combine two groups of concrete objects

Children are introduced to the concept of addition as combining two groups and counting them all to find the total. They are also introduced to the words 'how many' and 'altogether'.

You will need:

chalk; cards 1–5 from PCM 3; interlocking cubes in two different colours; double-sided (two-colour) counters (or dried lima/butter beans with one side painted); small pots or cups; selection of different small-world resources; small-world playmats and 'environments'

Getting started

- Say the action rhyme *Ten little fishies* (16), to practise counting forwards from 1 to 10.

Teaching

- Take the children outdoors. Draw two large circles on the ground using chalk. Ask 2 children to stand inside one circle and 1 child to stand in the other. Ask: *How many children are there in this circle? And how many in this circle?*
- Ask the single child to move into the circle with 2 children. Ask: *Now how many children are there altogether in this circle?* Emphasise the words 'how many' and 'altogether'.
- Take responses, then ask all the children to count with you to check: *1, 2, 3. Altogether there are 3 children in the circle. 2 and 1 makes 3.*
- Ask the 3 children in the circle to go back to the other children.
- Then ask 2 children to stand inside one circle and 3 children to stand in the other. Repeat as above.

Teaching notes

- Continue to combine groups of children that total 4 (3 and 1, 2 and 2), 5 (4 and 1) and 6 (4 and 2, 5 and 1, 3 and 3).

Explore in groups

Counters

- Provide small pots or cups, each containing 5 double-sided counters. Invite children to take a pot and tip the counters out. Ask them to say how many there are of each colour, and a corresponding addition statement. (Note: if the counters all land with the same colour facing up, ask children to tip the counters out again.)

Variation: put 4 or 6 double-sided counters in the pots.

Combining towers

- Shuffle two sets of cards 1–5 from PCM 3 (cubes). Place them face down in a pile. Also provide interlocking cubes in two different colours. Invite children to take a card, and build a tower using that number of cubes of one colour. Then ask them to take another card and repeat using the other colour cubes. They put both towers together to make one tall tower, then work out how many cubes altogether.

Counting collections station

- Provide 5 small-world resources of the same type, e.g. farm animals, and 2 playmats or 'environments', e.g. fields. Encourage children to discover all the different ways they can arrange the 5 resources into two groups.

Variation: provide groups of 4 and 6 small-world resources.

Session 2: Combine two groups of concrete objects and pictorial representations

Children consolidate their understanding of addition as combining two groups and counting them all to find the total. They are also introduced to the word 'total'.

You will need:

cards 1–5 from PCM 1; plastic or real fruit: lemon, apples, oranges, strawberries and bananas; 2 'fruit' boxes; basket; counters (or similar); small plates or dishes; glue; paper; coloured pencils or crayons, or paint, water, paintbrushes, aprons

Getting started

- Use the Number fluency activity *Show me (2)*, to practise counting groups of up to 10 objects.

Teaching

- Show children each of the fruit cards from PCM 1 in turn. Ask them to tell you how many there are on each card.
- Then take two of the cards, e.g. the cards with 4 strawberries and 2 apples. Place corresponding numbers of plastic or real fruit into the two boxes, e.g. 4 strawberries in one box and 2 apples in the other.
- Point and say: *There are 4 strawberries in this box and 2 apples in this box. Let's see how many pieces of fruit that is altogether.* Emphasise the words 'how many' and 'altogether'.
- One by one, take each of the 4 strawberries from the box and place them in the basket, counting: *1, 2, 3, 4*. Then take each of the 2 apples from the box and place them in the basket, counting: *... 5, 6*.
- Confirm the total number of fruit in the basket: *The total number of fruit is 6.* Emphasise the word 'total'.
- Repeat above, for two different fruit cards.

Explore in groups

Making 6 (adult-led)

- Give each child a set of fruit cards, and an additional 3 oranges card, from PCM 1. Ask the children to choose pairs of cards to make a total of 6 pieces of fruit. When complete, help children to glue the pairs of cards back-to-back.

Combining fruit

- Shuffle two sets of cards 1–5 from PCM 1 (fruit). Place them face down in a pile. Provide plastic fruit (or counters or similar), and two small plates or dishes. Encourage children to take a card and count out that number of fruit, placing them onto one of the plates. They take another card and repeat, placing the fruit on the other plate. Ask them to tell you how many pieces of fruit are on each plate, and how many altogether.

Painting and drawing station

- Invite children to draw or paint two boxes of fruit. Tell the children that altogether there must be 7 pieces of fruit in the two boxes.

Session 3: Combine two groups of objects in a picture

Children consolidate their understanding of addition as combining two groups and counting them all to find the total.

You will need:

Maths Foundation Reading Anthology; cards 1–5 from PCMs 1–3; paper (optional)

Getting started

- Use the Number fluency activity Finger counting (1), to practise counting forwards from 1 to 10.

Teaching

- Show pages 2 and 3 (At home) of the Reading Anthology. Discuss the picture with the children. At this stage do not focus on addition. Just ask children to count groups of up to 10 objects, e.g. How many things are on the sofa/small table/big table? How many things are in the picture/basket?
- Then move onto combining groups of objects. Ask questions such as: How many things are on the sofa? And how many things are on the small table? How many is this altogether? Only choose pairs of groups with totals no more than 10. Emphasise the words 'how many', 'altogether' and 'total'.

Explore in groups

How many fingers?

- Shuffle two sets of cards 1–5 from PCM 2 (fingers). Spread them out face up. Encourage children to choose two cards. Ask them to say how many fingers are on each card and how many fingers there are altogether.

Variation: use cards 1–5 from PCMs 1 (fruit) or 3 (cubes).

How many animals?

- Give each child or pair a copy of the Reading Anthology opened at page 10 (How many animals?). You may want to place a sheet of paper over page 11 so that children do not refer to it. Encourage children to take turns to point to two groups of animals, and say how many of each animal there are and how many there are altogether.

Maths Foundation Reading Anthology (adult-led)

- Together, look at pages 8 and 9 (The stream) of the Reading Anthology. Discuss the picture with the children. Point to different animals/objects in the picture. Ask children to tell you how many there are. Occasionally, ask questions such as: How many ducks and dragonflies are there? Altogether, how many snails and frogs can you see? Emphasise the words 'how many', 'altogether' and 'total'.

Session 4: Use pictures to record addition as combining two groups

Children consolidate their understanding of addition as combining two groups. They record the result using pictures. They are introduced to the word 'add'.

You will need:

Maths Foundation Reading Anthology; Tree Digital Tool; cards 1–5 from PCMs 1 and 3; PCM 6; paper; coloured pencils or crayons, or paint, water, paintbrushes, aprons

Getting started

- Show pages 8 and 9 (The stream) of the Reading Anthology. Discuss the picture with the children. Point to different animals/objects in the picture. Ask children to say how many there are. Occasionally, point and ask questions such as: How many ducks and dragonflies are there? Altogether, how many snails and frogs can you see? Emphasise the words 'how many', 'altogether' and 'total'.

Teaching notes

Teaching

- Display the Tree Digital Tool. Set it to show 2 trees. Place 3 apples onto the first tree and 2 apples onto the second tree. Point and say: *There are 3 apples on this tree, and 2 apples on this tree. How many apples are there altogether?*
- Take responses. Count all the apples on both trees. Confirm: *There are 5 apples. 3 add 2 is 5.* Emphasise the word 'add'.
- Click 'Clear all'. Repeat several times, placing up to 5 apples on each tree.
- Repeat again, several times, using other objects as well as apples. For example, 4 apples and 2 birds, or 3 birds and 5 ants. Each time ensure that totals are 10 or less.

Explore in groups

Apples trees

- Give each child two of the cards 1–5 from PCM 3 (cubes) and a copy of PCM 6. Children count the cubes on one card and draw that number of apples on the first tree. They then count the cubes on the second card and draw that number of apples on the second tree. When complete, ask each child to say how many apples are in each tree, and how many apples there are altogether.

Maths Foundation Reading Anthology

- Give each child a copy of the *Reading Anthology* opened at pages 8 and 9 (The stream). Tell each child two different animals, e.g. snails and ducks. Children count how many there are of each type of animal and draw or paint their own picture of the stream showing the same number of each animal. When complete, ask each child to say how many of each type of animal are in their picture, and how many animals there are altogether.

Fruit bowl

- Before the activity, prepare sheets of paper by drawing a 'fruit bowl' (large oval) on each. Shuffle a set of cards 1–5 from PCM 1 (fruit). Spread them out face up. Encourage children to choose two cards. Suggest that they draw the fruit shown on the cards in their fruit bowl (e.g. 5 bananas and 3 oranges). Ask children to say how many of each type of fruit are in the bowl, and how many fruit there are altogether.

Session 5: Use marks to record addition as combining two groups

Children consolidate their understanding of addition as combining two groups. They record the result using marks.

You will need:

Unit 7 Session 5 slide; PCMs 5, 9 and 14; 2 large 1 to 6 dot dice; counters; 2 small plates or dishes; paper; coloured pencils or crayons; small pots or cups; double-sided (two-colour) counters (or dried lima/butter beans with one side painted); non-transparent ('feely') bag containing 5 red and 5 blue interlocking cubes; *Maths Foundation Activity Book B*; pencils

Getting started

- Before the session, on two dice cover the 6-dot face with paper and draw the 3-dot pattern on it.
- Briefly hold up one of the subitising cards from PCM 9 (dots in regular formations). Ask children to say how many dots there. Show the card again and count.
- Repeat for the remaining 5 cards. Shuffle and repeat.

Teaching

- Roll one of the dice, e.g. 5. Ask: *How many dots?*
- Ask a child to count out that many counters and put them on a plate. Ask: *How many counters is this?*
- Represent this by drawing spots on the left side of the ladybird on slide 1. Say: *We can show these 5 counters by drawing 5 spots on the ladybird.*
- Now roll the second dice, e.g. 3. Repeat as above. Ask a child to put that number of counters on the second plate. Draw the corresponding number of spots on the right side of the ladybird.
- Point to each side of the ladybird: *The total of 5 add 3 is 8.*
- Hold up the two dice so children can see the numbers rolled: *5 add 3 is 8.* Emphasise the word 'add'.
- Point to the two plates: *5 add 3 makes 8.* Emphasise the word 'makes'.
- Remove the counters from the plates and wipe the spots from the slide.
- Repeat several times.

Explore in groups

Counters (recording marks)

- Provide sheets of paper folded in half, small pots or cups each containing 6 double-sided counters, and coloured pencils in two colours to match the colours of the counters. Invite children to take a pot and tip the counters out. Prompt them to represent the counters by drawing dots in one colour on one side of their paper, and dots in another colour on the other side.

Variation: put 4, 5 or 7 double-sided counters in each pot.

Ladybirds (recording marks)

- Shuffle two sets of the subitising cards from PCM 9 (dots in regular formations). Spread out the 10 cards face up. Provide ladybird cards from PCM 14. Invite children to choose a card and draw that number of spots on one side of a ladybird. They then choose another card and draw that number of spots on the other side.

Interlocking cubes (recording marks)

- Place the numeral 7 card from PCM 5 next to a non-transparent ('feely') bag containing 5 red and 5 blue interlocking cubes. Provide sheets of paper. Invite children to take 7 cubes from the bag without looking. Prompt them to arrange them into a blue tower and a red tower. Then ask them to join the two towers together. Suggest that they represent the total number of cubes on their paper.

Variation: provide numeral card 6, 8 or 9 instead of 7.

Maths Foundation Activity Book B (adult-led)

Page 6 – Add

When children have completed the page, ask them to say a corresponding addition statement for each pair, e.g. *5 add 3 makes 8.*

Session 6: Record addition as combining two groups using a ten-frame

Children consolidate their understanding of addition as combining two groups. They record the result using a ten-frame. They are introduced to the phrase 'is equal to'.

You will need:

Unit 7 Session 6 slide; PCMs 5, 9–11; 2 large 1 to 6 dot dice with 6-dot faces changed to 3-dot; counters; 2 small plates or dishes; small pots or cups; double-sided (two-colour) counters (or dried lima/butter beans with one side painted); coloured pencils or crayons; non-transparent ('feely') bag containing 5 red and 5 blue interlocking cubes; *Maths Foundation Activity Book B*

Getting started

- Briefly hold up one of the subitising cards from PCM 9 (dots in regular formations). Ask children to say how many dots there are. Show the card again and count.
- Repeat for the remaining 5 cards. Shuffle and repeat.
- Then repeat using the subitising cards from PCM 10 (dots in irregular formations).

Teaching

- Roll one of the dice, e.g. 4. Ask: *How many dots?*
- Ask a child to count out that many counters and put them on a plate. Ask: *How many counters is this?*
- Represent this by drawing 4 dots on the ten-frame on slide 1. Say: *We can show these 4 counters by drawing 4 dots on this ten-frame.*
- Now roll the second dice, e.g. 3. Repeat as above. Ask a child to put that number of counters on the second plate. Draw 3 dots in a different colour on the ten-frame.
- Point to each set of coloured dots on the ten-frame: *4 add 3 is equal to 7.* Emphasise the words 'add' and 'equal'. Explain: *The word 'equal' means 'is the same as'.*
- Hold up the two dice so children can see the numbers rolled: *4 add 3 is equal to 7.*
- Point to the two plates: *4 add 3 is equal to 7.*
- Remove the counters from the plates and wipe the dots from the slide.
- Repeat several times.

Teaching notes

Explore in groups

Counters (recording using ten-frames)

- Provide blank ten-frame cards from PCM 11, small pots or cups each containing 8 double-sided counters, and coloured pencils in two colours to match the colours of the counters. Invite children to take a pot and tip the counters out. Prompt them to represent the counters on a ten-frame.

Variation: put 6, 7 or 9 double-sided counters in each pot.

Subitising cards (recording using ten-frames)

- Shuffle two sets of cards from PCM 9. Spread them out face up. Provide blank ten-frame cards from PCM 11 and coloured pencils in two colours. Invite children to choose a card, and draw that number of dots in one colour on a ten-frame. Then ask them to choose another card and draw that number of dots on the ten-frame in another colour.

Interlocking cubes (recording using ten-frames)

- Place the numeral 9 card from PCM 5 next to a non-transparent ('feely') bag containing 5 red and 5 blue interlocking cubes. Provide blank ten-frame cards from PCM 11. Invite children to take 9 cubes from the bag without looking. Prompt them to arrange them into a red tower and a blue tower. Then ask them to join the two towers together. Suggest that they represent the total number of cubes on a ten-frame.

Variation: provide numeral card 6, 7 or 8 instead of 9.

Maths Foundation Activity Book B (adult-led)

Page 7 – Add

When children have completed the page, ask them to say a corresponding addition number sentence for each domino, e.g. *6 add 2 equals 8*.

Session 7: Record addition as combining two groups using numbers (A)

Children consolidate their understanding of addition as combining two groups. They record the result using numbers. They are introduced to the word 'plus'.

You will need:

PCMs 1, 2, 3, 5 and 15; Counting Digital Tool; small-world people; small pieces of paper; pencils; washing line (or string); 10 clothes pegs; 10 socks (or similar) in two different colours (5 of each); *Maths Foundation Activity Book B*

Getting started

- Use the Number fluency activity *Match the cards* (14), to practise matching groups of 1 to 10 objects with the corresponding number.

Teaching

- Display the Counting Digital Tool. Set it to show the pond scene with the frogs and fish.
- Place 3 frogs on the lily pads and 2 fish in the pond.
- Ask the children to count the frogs. Then point and say: *There are 3 frogs.* Click on the frog card to the right of the scene to reveal the number '3'.
- Ask the children to count the fish. Then point and say: *There are 2 fish.* Click on the fish card to reveal the number '2'.
- Then ask: *Altogether how many frogs and fish are there?* Emphasise the word 'altogether'.
- Discuss the responses. Count, pointing to each creature: *1, 2, 3, 4, 5. Altogether, there are 5 frogs and fish.* Click on the frog and fish card to reveal the total '5'.
- Point to the cards: *3 plus 2 is equal to 5.* Emphasise the words 'plus' and 'equal'.
- Click 'Clear all'. Click on the three cards to hide the numbers.
- Repeat several times. Make sure each total is 10 or less.

Explore in groups

Adding fruit

- Shuffle a set of cards 1–5 from PCM 1 (fruit). Spread them out face up. Also provide a set of numeral cards 1–10 from PCM 5. Encourage children to work in pairs. Each child chooses a fruit card and finds the matching numeral card. Prompt children to count the total number of fruit and find the corresponding numeral card.

Variation: use cards from PCM 2 (fingers) or 3 (cubes) instead of PCM 1.

Counting collections station

• Before the activity, draw a very simple house outline on a piece of paper. Draw a line across it to divide the house into 'upstairs' and 'downstairs'. Also provide 10 small-world people, a 1–10 number track from PCM 15, and small pieces of paper. Encourage children to place some people 'upstairs', and some people 'downstairs'. Prompt them to count the total number of people, and write the number on a small piece of paper. (They can look at the number track to see how to write the number.)

Washing line addition

• Before the activity, set up a washing line. Provide 10 pegs, 10 socks (or small pieces of fabric or card) in two different colours (5 of each), a 1–10 number track from PCM 15, and a pencil. Encourage children to work in pairs. One child chooses how many socks of one colour to peg on the line. The other child chooses how many socks of the other colour to peg on the line. They count the total number of socks together, and draw a dot on the matching number on the number track.

Maths Foundation Activity Book B (adult-led)

Page 8 – Add

When children have completed the page, ask them to say a corresponding addition number sentence for each box, e.g. *1 plus 4 is equal to 5.*

Session 8: Record addition as combining two groups using numbers (B)

Children consolidate their understanding of addition as combining two groups. They continue to record the result using numbers.

You will need:

PCMs 1, 5, 10 and 14; Unit 7 Session 8 slide; 2 large 1 to 6 dot dice with 6-dot faces changed to 3-dot; counters; 2 small plates or dishes; small pots or cups; double-sided (two-colour) counters (or dried lima/butter beans with one side painted); pencils; non-transparent ('feely') bag containing 5 red and 5 blue interlocking cubes; *Maths Foundation Activity Book B*

Getting started

• Use the Number fluency activity *Match the cards* (14), to practise matching groups of 1 to 10 objects with the corresponding number. (Alter the activity slightly by shuffling cards 1–10 from PCM 1 and placing them face down in a pile, and giving each child a numeral card 1–10 from PCM 5.)

Teaching

• Make sure each child keeps hold of their numeral card from *Getting started*.
• Roll one of the dice, e.g. 4. Ask: *How many dots?*
• Ask a child to count out that many counters and put them on a plate. Ask: *How many counters is this?*
• Ask all the children that have the numeral card that shows this number to hold it up. Say: *That's right. There are 4 counters and you are all holding up a card that shows the number 4.*
• Display slide 1. Write the number (e.g. 4) on the left side of the ladybird.
• Now roll the second dice, e.g. 3. Repeat as above. Ask a child to put that number of counters on the second plate, ask all the children with the corresponding numeral card to hold it up, and write the number (e.g. 3) on the right side of the ladybird.
• Point to each side of the ladybird: *4 add 3 is 7.* Emphasise the word 'add'. As you say '7' write the number 7 beside the ladybird.
• Hold up the two dice so children can see the numbers rolled: *4 plus 3 is equal to 7.* Emphasise the words 'plus' and 'equal'.
• Point to the two plates: *The total, of 4 and 3 is 7.* Emphasise the word 'total'.
• Remove the counters from the plates and wipe the numbers from the slide.
• Repeat several times.

Teaching notes

Explore in groups

Counters (recording using numbers)

- Provide ladybird cards from PCM 14 and pots or cups each containing 9 double-sided counters. Invite children to take a pot and tip the counters out. Prompt them to represent the counters on a ladybird card: they write the number of counters of one colour on one side of the ladybird, and the number of counters of the other colour on the other side. Ask them to write the total number of counters underneath the ladybird.

Variation: put 6, 7, 8 or 10 double-sided counters in each pot.

Subitising cards (recording using numbers)

- Shuffle two sets of subitising cards from PCM 10 (dots in irregular formations). Spread them out face up. Provide ladybird cards from PCM 14. Invite children to choose a card and write the number of dots on one side of a ladybird. Then ask them to choose another card and write that number of dots on the other side of the ladybird. Ask them to write the total number of dots underneath the ladybird.

Interlocking cubes (recording using numbers)

- Place the numeral 8 card from PCM 5 next to a non-transparent ('feely') bag containing 5 red and 5 blue interlocking cubes. Provide ladybird cards from PCM 14. Invite children to take 8 cubes from the bag without looking. Prompt them to arrange them into a red tower and a blue tower. Then ask them to join the two towers together. Suggest that they represent the cubes by writing the number of red cubes on one side of a ladybird and the number of blue cubes on the other side. They then write the total number of cubes underneath the ladybird.

Variation: provide numeral card 6, 7 or 9 instead of 8.

Maths Foundation Activity Book B (adult-led)

Page 9 – Add

When children have completed the page, ask them to say a corresponding addition number sentence for each ladybird, e.g. *5 plus 5 is equal to 10.*

Assessment opportunities

Assess children's learning against the objectives for this unit, using the guidance on formative assessment on pages 24–25, and record your observations in the Unit 7 progress tracking grid on page 29. The relevant pages of *Activity Book B* can also be used for assessment.

Can the children:

- count up to 10 objects?
- visually recognise a quantity of 6 or fewer: *subitising*?
- recognise and attempt to write (as numerals) numbers 1 to 10?
- demonstrate an understanding of addition as combining two sets?
- use the vocabulary associated with addition as combining two sets: 'how many', 'how much', 'altogether', 'total', 'add', 'plus', and 'is equal to'?
- begin to record an addition using pictures, marks or numbers?

Unit 8 Addition as counting on

Theme: Water

Overview

There are two different ways of looking at addition:

1) combining two quantities into a single quantity to work out the total – putting together by counting all

2) a given quantity is increased by another quantity – becoming greater by counting on.

This unit focuses on the second structure of addition: counting on. The key language in this unit includes: 'more', 'start at and count on' and 'start, then, now'. This unit also consolidates language associated with addition that was introduced in Unit 7: 'add', 'plus', 'total' and 'is equal to'.

Children apply their understanding of counting on to find 1 more than a number from 1 to 9.

Learning objectives

Number – Understanding addition and subtraction	• *4a In practical activities and discussions begin to use the vocabulary involved in addition: [combining two sets and]* **counting on.** • **4c Find 1 more [and 1 less] than a number from 1 to 10.**

Learning objectives in italics have been taught previously. In this unit they are consolidated and/or extended.

Learning objectives (or parts of objectives) in bold are taught for the first time in this unit.

[This part of the learning objective is not taught in this unit.]

Vocabulary

number, count, one, two, three, …, eight, nine, ten, start at, count forwards, count on, more, 1 more, start, then, now, total, add, plus, makes, is equal to

Making connections

English: Water

Science: Water

Preparation

You will need:

- *Maths Foundation Activity Book B*, pages 10–13.
- Digital Tools: Tree, Number track, Counting.
- Cards 1–9 from PCMs 1 and 3.
- Cards 1–5 from PCM 2.
- PCMs 5, 6 and 15.
- Number fluency games and activities: *Continue counting* (3) (page 198).
- Rhymes and songs: *One potato, two potatoes* (14), *Ten little fishies* (16), *One, two, buckle my shoe* (20) (pages 213–216).

Teaching notes

- Large 1–10 number bunting and/or large number track.
- Large and small 1 to 6 dot dice.
- 8 chairs.
- Beads and laces.
- Trays, small pots or cups, plates or dishes
- Small counters.
- '1-2' counters (counters with '1' written on one side, and '2' on the other).
- Chalk.
- Interlocking cubes.
- Pencils, coloured pencils.
- Beanbags.
- Large hoop.
- Counting collections station: a selection of water-themed small-world resources (e.g. sealife or boats), plates or dishes.

- Waterplay station: water tray, water-themed small-world resources.
- Sand station: sand tray/pit, moulds, flags, shells.
- Painting and drawing station: paper, paint, water, paintbrushes, aprons.

Before starting

Display number bunting or a large number track showing numerals 1–10. Each numeral should have a matching picture (e.g. 6 dots below the number 6). Refer to it regularly throughout the unit.

Before teaching this unit, look back at the Assessment section at the end of Unit 7 (page 94), to identify the prerequisite learning for this unit.

Common difficulties

To begin, children may re-count all of the objects in a group to see how many there are altogether, e.g. 1, 2, 3, 4, … 5, 6. However, they should be constantly encouraged to count on, e.g. start at 4 and count on: 5, 6.

Session 1: Count forwards, starting from any number to 10

Children count forwards, starting from, and ending at, any number from 1 to 10.

You will need:

large and small 1 to 6 dot dice; PCM 15; interlocking cubes; chalk

Getting started

- Ensure that all children can clearly see the large 1–10 number track/bunting.
- Say the rhyme *One, two, buckle my shoe* (20), practise counting forwards from 1 to 10 (stop at 'Nine, ten, a good, fat hen').

Teaching

- Point to '3' on the number track. Say: *We're going to count to 10 again, but this time we're not going to start from 1. We're going to count on from 3. Ready? Let's go!* Point to 3 on the number track and count on together along the track to 10.
- Repeat several times, counting on to 10 from a number other than 1. Occasionally, ask children to count on to a number other than 10, e.g. from 2 to 8.
- Without referring to the number track, occasionally ask questions such as: *What number comes after 4/9/3 …? What's the next number in this count: 3, 4, 5 …? Tell me the number that comes next: 5, 6, 7, 8 …*
- Show children a large 1 to 6 dot dice. Roll the dice. Ask children to count on from that number to 10 (or a number less than 10). Again, point to the numbers on the track as children count.
- Repeat several times.

Explore in groups

Repeat *Teaching* (adult-led)

- Repeat the *Teaching* activity with smaller groups of children. Give each child a 1–10 number track from PCM 15. Encourage them to count on, starting from any number from 1 to 10, pointing to the starting number on the track and counting on along the track to 10, or a number less than 10.

Number track counting

- Provide 1–10 number tracks from PCM 15 and small 1 to 6 dot dice. Invite children to roll the dice and place their finger on that number on the number track. Then ask them to count on along the track to 10.

Variation: draw a number track on the ground outdoors using chalk, give children a large 1 to 6 dot dice, and ask them to jump along the number track to match the dice rolls.

Counting on to 10 using cubes

- Provide interlocking cubes. Invite children to take a small number of cubes (e.g. 4) and build a tower. Then ask them to count on from that number to 10. As they say each number, they pick up one more cube and place it next to their tower. Ask them to check they have counted correctly: they say the number of cubes in the tower (e.g. *4*), then count on, touching the loose cubes one by one (e.g. *5, 6, 7, 8, 9, 10*).

Session 2: Addition as counting on (A)

Children are introduced to addition as counting on. They use the word 'more'.

You will need:

8 chairs; cards 1–5 from PCM 3; beads and laces; small pots or cups; plates or dishes; counters; selection of small-world resources

Getting started

- Use the Number fluency activity *Continue counting* (3), to practise counting forwards, starting from any number from 1 to 10.

Teaching

- Place 8 chairs to look like the seats in a large car: a row of 2, then two rows of 3 chairs behind them.
- Say: *This is our car! Some of you are going on a trip.* Ask 2 children to sit in the front seats. Say: *We start with 2 children in the car.* Emphasise the word 'start'.

- Then ask 3 children to sit in the next row. Ask: *How many more children are going on the trip?* Emphasise the word 'more'.
- Ask: *How many children are now in the car?* Emphasise the word 'now'.
- Point and say: *We started with 2 children in the car, then 3 more children joined them. Now there are 5 children in the car.* Emphasise the words 'started', 'then', 'more' and 'now'. *2 and 3 more makes 5.*
- Ask the 5 children to return to the others.
- Repeat several times for different groups of children up to a total of 8.

Explore in groups

More beads

- Shuffle two sets of cards 1–5 from PCM 3 (cubes). Place them face down in a pile. Provide a tray of beads and laces. Invite children to take a card (e.g. 5), count out that number of beads of one colour, and thread them onto a lace. Then ask them to take another card (e.g. 4), and add that number of beads to their lace, in a different colour. Prompt them to make a statement about their bead string, e.g. *5 and 4 more makes 9.*

Counting collections station

- Provide small-world resources of the same type, e.g. sealife, and plates or dishes. Encourage children to play in pairs. Ask each child to put up to 5 resources on their plate (e.g. 3). Each child then asks their partner for 'more', e.g. *'Can I have 2 more fish?'* Each child then takes that many more resources and says a corresponding statement, e.g. *3 and 2 more makes 5.*

Counters (adult-led)

- Provide pots each containing 10 counters, and plates/dishes. Ask children to play in pairs. Child A takes a pot of counters. Child B takes a plate. Child B says a number from 1 to 5, e.g. 3. Child A puts that number of counters onto the plate. Child B then says, e.g. *Give me 4 more please.* Child A then puts that many more counters onto the plate. Both children then say how many counters there are altogether on the plate.

Teaching notes

Session 3: Find 1 more than a number from 1 to 9

Children consolidate their understanding of addition as counting on. They use a number track as a tool for finding 1 more than a given number.

You will need:

large floor 1–10 number track (if necessary, use chalk to draw a track outdoors); large 1 to 6 dot dice; water tray; small-world resources; cards 1–9 from PCMs 3 and 5; PCM 15; interlocking cubes; *Maths Foundation Activity Book B*; pencils

Getting started

- Ensure that all children can see the large floor 1–10 number track.
- Say the action rhyme *Ten little fishies* (16), to practise counting forwards from 1 to 10.

Teaching

- Roll the dice, e.g. 5. Ask: *How many dots are there?* Stand on that number on the number track.
- Say: *Now I'm going to jump on 1 more space along the track. What number will I land on?*
- Jump on 1 space. Say: *I have landed on 6. 6 is 1 more than 5.* Emphasise the words '1 more'.
- Repeat two or three times. Model the different ways the vocabulary can be used, e.g. *4 and 1 more is 5. 1 more than 2 is 3.*
- Ask a child to roll the dice, stand on that number on the number track and then jump on 1 space along the track. Encourage them to use the words '1 more'.
- Repeat several times with different children.
- Ask children to say the number that is 1 more than 7, 8 or 9.

Explore in groups

Waterplay station

- Place a selection of water-themed small-world resources, such as sealife or boats, next to the water tray. Encourage children to play in pairs. One child says a number from 1 to 9, e.g. 7. The other child then puts 1 more than that number of resources in the water, and says e.g. *8 is 1 more than 7.*

1 more

- Shuffle a set of numeral cards 1–9 from PCM 5. Place them face down in a pile. Encourage children to play a game in pairs. They turn over the top card, and both say the number that is 1 more. The first child to say the correct answer keeps the card. The overall winner is the child with the most cards.

1 more than

- Shuffle a set of cards 1–9 from PCM 3 (cubes). Place them face down in a pile. Provide a tray of interlocking cubes, and a 1–10 number track from PCM 15. Invite children to take a card. Ask them to build a tower with 1 more cube than the number on the card. Ask them to point to that number on the number track and say e.g. *1 more than 7 is 8.*

Maths Foundation Activity Book B (adult-led)

Page 10 – 1 more

Session 4: Addition as counting on along a number track

Children consolidate their understanding of addition as counting on. They use a number track as a tool for counting on.

You will need:

large floor 1–10 number track (if necessary, use chalk to draw a track outdoors); large 1 to 6 dot dice; cards 1–5 from PCMs 3; PCM 15; small pots or cups; small counters; '1-2' counters (counters with '1' written on one side, and '2' on the other); chalk

Getting started

- Before the session, on the dice cover the 6-dot face with paper and draw the 3-dot pattern on it.
- Ensure that all children can see the large floor 1–10 number track.
- Use the Number fluency activity *Continue counting* (3), to practise counting forwards, starting from any number from 1 to 10.

Teaching

- Roll the dice, e.g. 3. Ask: *How many dots are there?* Stand on that number on the number track.
- Roll the dice again, e.g. 4. Ask: *Now how many dots are there?* Jump along the number track from 3 to 7, counting on 4 as you jump: *1, 2, 3, 4.*
- Say: *I started at 3, then I counted on 4. I ended on 7.* Emphasise the words 'started', 'then' and 'counted on'. *3 and 4 more make 7.* Emphasise the word 'more'.
- Repeat several times.
- Roll the dice, and stand on that number on the number track. Roll the dice again. Ask the child to take that many jumps along the number track. As the child does this ask the rest of the children to count on.
- Repeat several times with different children.

Explore in groups

Number track count

- Shuffle two sets of cards 1–5 from PCM 3 (cubes). Place them face down in a pile. Provide a pot of small counters in two different colours and 1–10 number tracks from PCM 15. Invite children to take the top card, e.g. 3. Ask them to count out that number of counters of one colour, and place them on the number track starting from 1. Ask them to take another card, e.g. 3. They take that number of counters of the other colour, and place them on the number track. Prompt children to make a corresponding statement, e.g. *3 and 3 more makes 6.*

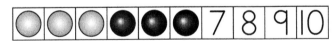

Jumping along the number track

- Before the activity, draw a large 1–10 number track on the ground outdoors using chalk. Encourage children to play in groups. They take turns to be the 'jumper'. A child calls out a starting number up to 5, e.g. 4. The 'jumper' jumps (or hops on one foot) along the number track until they reach the number. Another child calls out a second number up to 5, e.g. 2. The 'jumper' jumps that number of squares along the track. All the children together call out the final number, e.g. 7.

1 or 2 more (adult-led)

- Ask children to play a game in pairs. Give each child a 1–10 number track from PCM 15 and a small counter, and each pair a small pot or cup containing a '1-2' counter. Each child puts their counter on 1 on their number track. They take turns to tip the 1-2 counter out of the pot and move that number of spaces along their track. Prompt them to say a corresponding addition statement, e.g. *4 and 2 more makes 6.* The winner is the first child to reach the end of their track.

Session 5: Addition as counting on (B)

Children consolidate their understanding of addition as counting on. They use the words 'more' and 'add'.

You will need:

large 1–10 number track; trays of interlocking cubes; sand tray, moulds, pebbles, shells; cards 1–5 from PCM 3; paper; paint, water, paintbrushes, aprons

Getting started

- Ensure that all children can see the large 1–10 number track.
- Say the rhyme *One potato, two potatoes* (14), to practise counting on from 1 to 10 and to reinforce the word 'more'.

Teaching

- Say a number less than 5, e.g. 3. Point to it on the number track. Take that number of interlocking cubes, counting each one as you pick it up. Create a cube tower. Say: *I have 3 cubes.*
- Then say another number less than 5, e.g. 2. Say: *Now I want to add 2 more cubes to my tower.* Take that number of cubes, counting each one as you pick it up. Add them to the tower.
- Say: *I started with 3 cubes* [point to 3 on the number track], *then I added 2 more cubes* [from 3 count on 2 along the number track]. *Now I have 5 cubes* [point to 5]. Emphasise the words 'started', 'then', 'added', 'more' and 'now'.
- Repeat several times.

Teaching notes

- Ask a child to say a number less than 5 and make a matching cube tower. Ask another child to say a number less than 5. The first child adds that many cubes to their tower. During each stage of the process reinforce the key mathematical vocabulary and refer to the number track.
- Repeat several times with different children.

Explore in groups

Sand station

- Provide small moulds, and beach- or water-themed resources (e.g. shells, pebbles). Encourage children to play in pairs. One child says a number from 1 to 5, e.g. 4, and either makes that number of sandcastles, or sets out that number of resources. The other child then says another number from 1 to 5, e.g. 5, and adds that number of sandcastles/resources. Prompt them to say a corresponding statement, e.g. *4 and 5 more makes 9.*

Cube towers

- Shuffle two sets of cards 1–5 from PCM 3 (cubes). Place them face down in a pile. Provide a tray of interlocking cubes in two different colours. Invite children to take the top card, e.g. 3. Ask them to count out that number of cubes of one colour and make a tower. They then choose another card, e.g. 4, and add that number of cubes to their tower in a different colour. Prompt them to make a statement about their tower, e.g. *3 and 4 more makes 7.*

Painting and drawing station

- Let children paint a picture linked to the theme 'water'. Explain that you would like them to include two groups of things, e.g. red fish and blue fish. Ask them to start by painting one group. Ask questions to draw out the starting number, how many more are added, and how many there are in total (ensure that totals are 10 or less), e.g. *How many red fish are there? How many blue fish will you paint now? How many is that altogether?*

Session 6: Addition as counting on (C)

Children consolidate their understanding of addition as counting on. They use the words 'more', 'add', 'plus' and 'is equal to'.

You will need:

large 1–10 number track; Tree Digital Tool; cards 1–5 from PCM 1; PCM 6; coloured pencils; water-themed small-world resources; plate; *Maths Foundation Activity Book B*; pencils

Getting started

- Ensure that all the children can see the large 1–10 number track.
- Say the rhyme; *One potato, two potatoes* (14), to practise counting on from 1 to 10 and to reinforce the word 'more'.

Teaching

- Display the Tree Digital Tool. Set it to show 1 tree. Place 4 birds onto the tree, in a group near to each other. Ask: *How many birds are in the tree?*
- Place 2 more birds onto the tree, in a separate group from the first 4. Say: *2 more birds fly into the tree. Now how many birds are in the tree?*
- Point to the two groups of birds and say: *4 and 2 more makes 6.* Emphasise the word 'more'. Count the 6 birds with the children. *4 add 2 is equal to 6. 4 plus 2 makes 6.* Emphasise the words 'add', 'equal' and 'plus'.
- Say: *We started with 4 birds* [point to 4 on the number track], *then 2 more birds flew into the tree* [from 4 count on 2 along the track]. *Now there are 6 birds in the tree* [point to 6]. Emphasise the words 'started', 'then', 'more' and 'now'.
- Click 'Clear all'. Repeat several times.

Explore in groups

More birds

- Before the activity, prepare copies of PCM 6. On each sheet, write a number less than 9 above the first tree, e.g. 6. Invite children to draw the corresponding number of birds on their first tree. Then ask, e.g. *How many birds have you drawn? Can you add 3 more to that number?* (Ensure that totals are 10 or less.) Ask children to draw the corresponding number of birds on the second tree, e.g. 9.

More fruit

- Shuffle two sets of cards 1–5 from PCM 1 (fruit). Place them face down in a pile. Invite children to play in pairs. The first child turns over a card, e.g. 4. Then the other child turns over a card, e.g. 2. Ask the children to say a corresponding addition statement using the language of counting on, e.g. *4 fruit and 2 more fruit makes 6 fruit.* Children repeat until all 10 cards have been used.

Counting collections station (adult-led)

- Place 10 water-themed small-world resources on a plate. Ask a child to take some (but not all) of the resources and count them out onto the table. Depending on the number of resources chosen, ask the children to tell you 1, 2, 3, … 9 more. (Ensure that totals are 10 or less.) If needed, children can check by counting out the additional resources. Repeat with different children choosing the resources.

Maths Foundation Activity Book B (adult-led)

Page 11 – 2 more

Encourage children to use the number track on the page to count on. They could record the totals either by drawing marks, or by attempting to write the numerals. When children have completed the page, ask them to say a corresponding addition statement for each cloud, e.g. *4 and 2 more makes 6.*

Session 7: Record addition as counting on using a number track (A)

Children consolidate their understanding of addition as counting on. They record the result using a number track.

You will need:

Number track Digital Tool; beanbags; large hoop; large 1 to 6 dot dice with 6-dot face changed to 3-dot; cards 1–5 from PCM 2; PCM 15; small counters; chalk; *Maths Foundation Activity Book B*; coloured pencils

Getting started

- Display the Number track Digital Tool. Set it to show a 1–10 number track.
- Use the Number fluency activity *Continue counting* (3), to practise counting forwards, starting from any number from 1 to 10.

Teaching

- Place a hoop on the floor.
- Roll the dice, e.g. 4. Ask: *How many dots are there?* Place that number of beanbags in the hoop.
- On the Number track Digital Tool, move the kangaroo to 4. Draw a circle around number 4 on the track. Say: *We start with 4 beanbags.*
- Now roll the dice again, e.g. 3. Ask: *Now how many dots are there?* Pick up 3 beanbags and show them to the children. Say: *Then we add 3 beanbags.*
- Ask: *How many beanbags are already in the hoop?* (4) *Let's add these 3, counting on from 4.* As you place each beanbag in the hoop say: *5, 6, 7. There are now 7 beanbags in the hoop.*
- Point to the kangaroo on the number track. Say: *We started with 4 beanbags, then we added 3 more: 1* [draw a jump from 4 to 5], *2* [draw a jump from 5 to 6], *3* [draw a jump from 6 to 7]. *Now there are 7 beanbags in the hoop.* Unclick 'Draw' and then click on 7 to highlight the answer on the track.

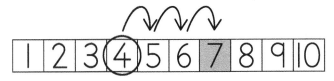

- Say: *So, 4* [point to 4 on the number track] *and 3 more* [point to the 3 jumps from 4 to 7] *make 7* [point to 7]. *How else can we say this?* Praise children who suggest other statements such as: *4 add 3 is equal to 7; 4 plus 3 equals 7; the total of 4 and 3 is 7.*
- Reset the Number track Digital Tool and remove the beanbags from the hoop. Repeat several times.

Explore in groups

Counting on fingers

- Shuffle a set of cards 1–5 from PCM 2 (fingers). Place them face down in a pile. Provide small counters and 1–10 number tracks from PCM 15 with a circle around number '2'. Encourage children to place a

Teaching notes

counter on 2 on a number track, and take a finger card from the pile. Ask them to count along the number track the number of fingers shown on the card. Prompt them to say a corresponding statement, e.g. *2 and 3 more makes 5.*

Variation: provide number tracks with number 1 (or 3, 4 or 5) circled.

Counting on dice

- Provide 1–10 number tracks from PCM 15, and the 1 to 6 dot dice with the 6-dot face changed to the 3-dot pattern. Encourage children to play as a group. They take turns to roll the dice and say the number, e.g. 5. Each of the other children puts their finger on that number on their own number track. The first child rolls the dice again and says the number followed by the word 'more', e.g. '3 more'. The other children count on along their number track that number of jumps and say the answer. e.g. 8.

Outdoor play

- Before the activity, draw a large 1–10 number track on the ground outdoors using chalk. Provide 1 beanbag in one colour (e.g. red) and 5 beanbags in a different colour (e.g. blue). Encourage children to play in groups. One child throws the red beanbag onto the number track, e.g. it lands on 3. (If it misses the track, or lands beyond 5, they can try again, or move it to the nearest number.) Another child says a number up to 5, e.g. 4. A third child collects that number of blue beanbags, and runs to places one on each square after the red beanbag. The children all say the total, e.g. 7. Prompt them to say a corresponding addition statement, e.g. *3 and 4 more makes 7.*

Maths Foundation Activity Book B (adult-led)

Page 12 – Add on

When children have completed the page, ask them to say a corresponding addition number sentence for each number track, e.g. *2 add 4 is equal to 6.*

Session 8: Record addition as counting on using a number track (B)

Children consolidate their understanding of addition as counting on. They record the result using a number track.

You will need:

large 1–10 number track; Counting Digital Tool; numeral cards 2–5 from PCM 5; small pots or cups; small counters; '1-2' counters (counters with '1' written on one side, and '2' on the other); 1 to 6 dot dice; PCM 15; water-themed small-world resources; *Maths Foundation Activity Book B*; coloured pencils

Getting started

- Ensure that all the children can see the large 1–10 number track.
- Use the Number fluency activity *Continue counting* (3), to practise counting forwards, starting from any number from 1 to 10.

Teaching

- Display the Counting Digital Tool. Set it to show the beach scene with the crab and fish.
- Slowly place 3 fish in a group on the left of the scene. Ask children to count them. Point to this number on the number track.
- Then place 4 fish in a second group on the right of the scene. Say: *We started with 3 fish. Then 4 more fish joined them. Now how many fish are there?*
- Point to the two groups of fish. Say: *3 and 4 more makes 7.* Emphasise the word 'more'.
- Count the 7 fish with the children. *3 add 4 is equal to 7. The total number of fish is 7.* Emphasise the words 'add', 'equal' and 'total'.
- Say: *We started with 3 fish* [point to 3 on the number track]*, then 4 more fish joined them* [from 3 count on 4 along the number track]*. Now there are 7 fish.* [point to 7]. Emphasise the words 'started', 'then', 'more' and 'now'.
- Click on the fish card to the right of the scene to reveal the number '7'.
- Click 'Clear all'. Click the fish card to hide the numeral 0. Repeat several times.

Explore in groups

1 or 2 more

- Shuffle 2 sets of numeral cards 2–5 from PCM 5. Place them face down in a pile. Provide small counters, 1–10 number tracks from PCM 15, and small pots each containing a '1-2' counter. Encourage children to play a game in pairs. Each child chooses a number card and puts their counter on that number on their number track. They then take turns to tip the 1-2 counter out of the pot and move their counter that number of spaces along the track. Prompt children to say a corresponding addition statement. The winner is the child whose total is closer to 10. This child keeps both number cards. If both children land on the same number, they each keep their card. The game continues until all 8 cards have been used. The overall winner is the child with the most cards.

Small-world counting on

- Provide 1–10 number tracks from PCM 15, water-themed small-world resources, and the 1 to 6 dot dice with the 6-dot face changed to the 6-dot pattern. Encourage children to play a game in pairs. They take turns to roll the dice. They place the matching number of small-world resources on their number track, covering the numbers. They then each roll the dice again, and add that number of resources onto their number track. Prompt them to say a corresponding addition statement. The winner is the child whose total is closer to 10.

Maths Foundation Activity Book B (adult-led)

Page 13 – Add more

When children have completed the page, ask them to say a corresponding addition statement for each picture, e.g. *3 and 4 more makes 7*.

Assessment opportunities

Assess children's learning against the objectives for this unit, using the guidance on formative assessment on pages 24–25, and record your observations in the Unit 8 progress tracking grid on page 30. The relevant pages of *Activity Book B* can also be used for assessment.

Can the children:

- count forwards, starting from, and ending at, any number from 1 to 10?
- count up to 10 objects?
- visually recognise a quantity of 6 or fewer: *subitising*?
- recognise and attempt to write (as numerals) numbers 1 to 10?
- demonstrate an understanding of addition as counting on, including using a number track?
- find 1 more than a number from 1 to 9?
- use the vocabulary associated with addition as counting on: 'more', 'total', 'add', 'plus' and 'is equal to'?
- begin to record an addition using pictures, marks or numbers?

Unit 9 Patterns and data

Theme: Growing and changing

Overview

This unit will develop children's awareness of patterns. They will observe, copy, continue and create patterns made of objects, pictures, sounds and actions. Children start with simple alternating AB patterns (e.g. square, triangle, square, triangle). They progress to different 'units of repeat': ABC (e.g. square, triangle, circle, square, triangle, circle) and ABB (e.g. square, triangle, triangle, square, triangle, triangle).

This unit also focuses on sorting, representing and describing data. In other units, children have compared, matched and sorted numbers, shapes and objects. In this unit they compare, match and sort a range of different objects using criteria such as shape, colour, size and type.

Seeking and exploring patterns, and sorting and classifying, are at the heart of mathematics. They provide opportunities for children to observe and verbalise generalisations. After teaching this unit, for the remainder of the year, try to find opportunities to repeat the unit's *Explore in groups* activities.

Learning objectives

Number – Patterns and sequences	• *5a Talk about, recognise and make simple patterns using concrete materials or pictorial representations.*
Statistics	• *10a Sort, represent and describe data using concrete materials or pictorial representations.*

Learning objectives in italics have been taught previously. In this unit they are consolidated and/or extended.

Vocabulary

size, big, small, shape, colour, pattern, repeat, repeating pattern, unit of repeat, next, match, sort, group, set, same, different, type

Making connections

English: Growing and changing

Science: Growing and changing

Preparation

You will need:

- *Maths Foundation Reading Anthology*, pages 14–15, 30 and 31.
- *Maths Foundation Activity Book B*, pages 14–17.
- Digital Tool: Beads and laces.

- PCMs 23 and 24 (preferably enlarged to A3).
- Rhymes and songs: *Clap your hands* (27) (page 218).
- Magazines, comics, catalogues, etc.

- Selection of objects suitable for pattern-making and sorting, e.g. beads and laces, 2D shapes (circles, triangles, squares, rectangles), pegboards and pegs, different-coloured interlocking cubes, different-coloured counters, building blocks, playdough and cutters, natural materials (leaves, sticks, stones, pine cones), small-world resources (e.g. animals and sealife) in different colours and sizes, buttons, Home corner props, soft toys.
- Selection of resources suitable for pattern-printing, e.g. stickers, self-inking stamps, plastic 3D shapes (cylinder, pyramid, cube, cuboid), stamp pads, paint, aprons.
- Coloured pencils and/or crayons and/or felt-tip pens.
- Chime bars/xylophone.
- Sorting hoops.
- Sorting trays.

Before starting

As part of Session 1, start to create a wall and table display of patterns and sorted objects. Throughout the unit ask children to look for examples of patterns (including fabric, wrapping paper, wallpaper samples, etc.) and objects that have been sorted, in magazines, comics, catalogues, etc. to include in the display.

Children will get the most out of this unit if they already have some experience of comparing, matching and sorting objects. This could be in a range of contexts, including shape and measures, as well as everyday classroom experiences such as tidying up. It is important to provide such experiences throughout the course of the year.

Maths background

The part of a pattern that is repeated is called the *unit of repeat.* To help children recognise the sequence in a repeating pattern, it is important to provide at least three full units of repeat, e.g. AB AB AB. When children continue a pattern, ask them to create at least two full units of repeat, e.g. AB AB AB AB AB.

Encourage children to say the pattern aloud. This helps them to identify the unit of repeat and supports them to continue the pattern.

Getting started

- Before the session, place some objects suitable for pattern-making on a table. Include a range of objects that can be used to create AB patterns in different contexts, e.g. shape, colour, size, type.
- Say the action rhyme *Clap your hands* (27). At the end of the rhyme, perform an AB pattern of claps, e.g. clap high above your head and then down by your knees (repeated three times). Repeat several times, with children joining in.

Teaching

- Choose some objects from the table, e.g. interlocking cubes in two different colours. Say: *Watch carefully, and tell me what you notice*. Make an AB pattern with the cubes, placing them side by side. Say the colour as you put down each cube, e.g. *red, blue, red, blue, red, blue*. Ask: *What do you notice?*
- Discuss responses. Say: *We call this a repeating pattern. This is a pattern of cubes where two colours, red and blue, repeat.*
- Show children the unit of repeat. Pick up each 'unit' (a red cube and a blue cube) one by one. Say: *This is a red-blue pattern. These two*

Session 1: AB patterns using objects

Children observe, copy, continue and create AB repeating patterns made from objects.

You will need:

selection of objects suitable for pattern-making, e.g. beads and laces, 2D shapes (circles, triangles, squares, rectangles), pegboards and pegs, interlocking cubes, counters, building blocks, playdough and cutters, natural materials (leaves, sticks, stones, pine cones), small-world resources in different colours and sizes

Teaching notes

colours keep repeating. See: here is another red-blue pattern, and another red-blue pattern.

- Repeat above, making a different AB pattern using cubes in two colours. Make sure to show three full units of repeat.
- Repeat a third time. Then ask: *If I wanted to continue this pattern, what colour cube would I put next? What would be the next colour after that? And the next?* Continue until the pattern has at least 10 cubes.
- Repeat above, but this time make an error, e.g. green, yellow, green, yellow, green, green. Ask: *Is this right? What have I done wrong? What colour should this cube be?*

Explore in groups

Pattern-making (using objects)

- Provide a range of resources that children can use to create patterns based on shape, colour, size and type (e.g. beads and laces, interlocking cubes, playdough, 2D shapes, counters, natural materials, pegboards and pegs, building blocks, small-world resources). Encourage children to make AB repeating patterns. Ask them to tell you about each pattern. Help them to identify the unit of repeat. You could continue a child's pattern, making a mistake; do they correct it?

Copy my pattern

- Provide beads and laces. Invite children to play in pairs. Each child threads beads onto a lace to create an AB repeating pattern. When they have created at least three units of repeat, ask children to take a new lace each, and copy each other's pattern.

Continue my pattern

- Provide interlocking cubes. Invite children to play in pairs. Each child makes a cube tower with an AB repeating pattern. When they have created at least three units of repeat, ask children to swap towers and continue each other's pattern. If needed, prompt them to continue for at least two units of repeat.

Session 2: AB patterns involving objects, pictures, sounds and actions

Children observe, copy, continue and create AB repeating patterns. They use objects, pictures, sounds and actions.

You will need:

Maths Foundation Reading Anthology; Beads and laces Digital Tool; selection of resources suitable for pattern-printing, e.g. stickers, self-inking stamps, plastic 3D shapes (cylinder, pyramid, cube, cuboid), stamp pads; coloured pencils, crayons and felt-tip pens; paint, water, aprons; PCM 23; chime bars/xylophone; *Maths Foundation Activity Book B*

Getting started

- Before the session, place a selection of resources suitable for pattern-printing on a table. Include a range of resources that can be used to create patterns based on shape, colour, size and type.
- Show pages 14 and 15 (Carpet patterns) of the *Reading Anthology*. Discuss the carpets with the children. Focus on the horizontal AB patterns (there are three different patterns on page 14 and two different patterns on page 15). Ask children to identify and describe each pattern, and say how it would continue. Point out the unit of repeat in each pattern.

Teaching

- Display the Beads and laces Digital Tool. Show 1 lace. Use blue and red beads to make an AB pattern. Show three units of repeat.
- Ask children to identify the pattern (e.g. *It goes: blue, red, blue, red, blue, red*). Point out the unit of repeat, e.g. *The part that repeats is one blue bead and one red bead. What bead comes next?*
- With children's help, continue the pattern for at least two units of repeat.
- Click 'Clear all'. Ask a child to come and choose two different beads to start a pattern, e.g. yellow and green. Say: *This is the part of a pattern that will repeat.* Add two more units of repeat. Ask another child to come and continue the pattern.
- Demonstrate a sound or action AB pattern, e.g. clap, stomp, clap, stomp, clap, stomp.

Say the pattern aloud and encourage children to join in. Point out the unit of repeat.

- Repeat for other sound and action AB patterns, e.g. tap shoulders, tap knees; arms up, arms down; sit, stand; jump, step; music patterns (using chime bars/xylophone); numbers (e.g. *one, two)*; phonic sounds (e.g. *ch, sh*).
- Ask a child to make their own sound or action AB pattern. Make sure they include three units of repeat. Ask other children to copy and continue the pattern.
- Repeat with one or two more children.

Explore in groups

Pattern-making (using objects)

- Repeat the *Pattern-making (using objects)* activity from Session 1 (page 106). See Session 1 'You will need' list for suggested resources.

Pattern-making (using pictures)

- Provide pattern tracks from PCM 23 and a range of resources that children can use to create patterns by printing (e.g. stickers, self-inking stamps, plastic 3D shapes (cylinder, pyramid, cube, cuboid), paint, stamp pads). Invite children to create AB repeating patterns. Ask them to tell you about each pattern. Help them to identify the unit of repeat. Turn a pattern 90°, then 180°, then 270°. Ask: *What do you notice? What is the same? What is different?*

Copy and continue my pattern

- Provide pattern tracks from PCM 23, and coloured pencils, crayons or felt-tip pens. Invite children to play in pairs. Each child draws an AB repeating pattern on a track. When they have drawn at least three units of repeat, ask children to take a new track each, and copy and continue each other's pattern.

Sound and action AB patterns (adult-led)

- Repeat the last part of the *Teaching* activity with smaller groups of children (ask children to copy, continue and create sound and action AB patterns). This could be done outdoors.

Maths Foundation Activity Book B (adult-led)

Page 14 – Patterns

When children have completed the page, help them to rotate it 90°, then 180°, then 270°. Ask: *What do you notice? What is the same? What is different?*

Session 3: ABC patterns using objects

Children observe, copy, continue and create ABC repeating patterns made from objects.

You will need:

selection of objects suitable for pattern-making, e.g. beads and laces, 2D shapes (circles, triangles, squares, rectangles), pegboards and pegs, interlocking cubes, counters, building blocks, playdough and cutters, natural materials (leaves, sticks, stones, pine cones), small-world resources in different colours and sizes

Getting started

- Before the session, place some objects suitable for pattern-making on a table. Include a range of objects that can be used to create ABC patterns in different contexts, e.g. shape, colour, size, type.
- Say the action rhyme *Clap your hands* (27). At the end of the rhyme, perform an ABC pattern of claps, e.g. clap to your left, then to your right, then high above your head (repeated three times). Repeat several times, with children joining in.

Teaching

- Choose some objects from the table, e.g. 2D shapes. Say: *Watch carefully, and tell me what you notice.* Make an ABC pattern using three shapes. Say the shape names as you place each one in the row, e.g. *circle, triangle, square.* Show at least three units of repeat. Ask: *What do you notice?*
- Discuss responses. Say: *This is a repeating pattern of three different shapes: a circle, a triangle and a square.*
- Show children the unit of repeat. Indicate each 'unit' (a circle, a triangle and a square) with your hands. Say: *This is a circle-triangle-square pattern. These three shapes keep repeating. See: here is another circle-triangle-square pattern.*

Teaching notes

- Repeat above, making a different ABC pattern using three shapes. Make sure to show at three full units of repeat.
- Repeat a third time. Then ask: *If I wanted to continue this pattern, what shape would I put next? What would be the next shape after that? And the next?* Continue until the pattern has at least 15 shapes.
- Repeat above, but this time make an error, e.g. circle, triangle, rectangle, circle, triangle, rectangle, rectangle, triangle, rectangle. Ask: *Is this right? What have I done wrong? What shape should be here instead?*

Explore in groups

Pattern-making (using objects)

- Repeat the *Pattern-making (using objects)* activity from Session 1 (page 106), but ask children to create ABC patterns. See Session 1 'You will need' list for suggested resources.

Copy my pattern

- Provide 2D shapes. Invite children to play in pairs. Each child creates an ABC repeating pattern of shapes. When they have created at least three units of repeat, ask children to copy each other's pattern.

Continue my pattern

- Take children outdoors, and ask them to create a collection of natural materials (e.g. leaves, sticks, stones, pine cones). Then invite children to play in pairs. Each child creates an ABC repeating pattern using the natural materials. When they have created at least three units of repeat, ask children to swap places and continue each other's pattern. If needed, prompt them to continue for at least two units of repeat.

Session 4: AB and ABC patterns involving objects, pictures, sounds and actions

Children observe, copy, continue and create AB and ABC repeating patterns. They use objects, pictures, sounds and actions.

You will need:

Maths Foundation Reading Anthology; Beads and laces Digital Tool; selection of resources suitable for pattern-printing, e.g. stickers, self-inking stamps, plastic 3D shapes (cylinder, pyramid, cube, cuboid), stamp pads; coloured pencils, crayons and felt-tip pens; paint, water, aprons; PCM 24; chime bars/xylophone; *Maths Foundation Activity Book B*

Getting started

- Before the session, place a selection of resources suitable for pattern-printing on a table. Include a range of resources that can be used to create patterns based on shape, colour, size and type.
- Show page 14 (Carpet patterns) (<u>not</u> page 15) of the *Reading Anthology*. Briefly remind children of the three different horizontal AB patterns. Then focus on the two different vertical ABC patterns. Ask children to identify and describe each pattern, and say how it would continue. Point out the unit of repeat in each pattern.

Teaching

- Display the Beads and laces Digital Tool. Show 1 lace. Use blue, red and yellow beads to make an ABC pattern. Show three units of repeat.
- Ask children to identify the pattern (e.g. *It goes blue, red, yellow, blue, red, yellow*). Point out the unit of repeat, e.g. *The part that repeats is one blue bead, one red bead and one yellow bead. What bead comes next?*
- With children's help, continue the pattern for at least two units of repeat.
- Click 'Clear all'. Ask a child to come and choose three beads to start a pattern, e.g. green, yellow, blue. Say: *This is the part of the pattern that will repeat.* Add two more units of repeat. Ask another child to come and continue the pattern.
- Demonstrate a sound or action ABC pattern, e.g. clap, stomp, click fingers, clap, stomp, click fingers, clap, stomp, click fingers. Say the pattern aloud and encourage children to join in. Point out the unit of repeat.
- Repeat for other sound and action ABC patterns, e.g. tap head, tap shoulders, tap knees; sit, stand, jump; jump, step, hop; music patterns (using chime bars/xylophone); numbers (e.g. *one, two, three*); phonic sounds (e.g. *ch, sh, th*).
- Ask a child to make their own sound or action ABC pattern. Make sure they include three

units of repeat. Ask other children to copy and continue the pattern.

- Repeat with one or two more children.

Explore in groups

Pattern-making (using objects)

- Repeat the *Pattern-making (using objects)* activity from Session 1 (page 106), but ask children to create either AB or ABC patterns. See Session 1 'You will need' list for suggested resources.

Pattern-making (using pictures)

- Provide pattern tracks from PCM 24 and a range of resources that children can use to create patterns by printing (e.g. stickers, self-inking stamps, plastic 3D shapes (cylinder, pyramid cube, cuboid), paint, stamp pads). Invite children to create AB or ABC repeating patterns. Ask them to tell you about each pattern. Help them to identify the unit of repeat. Turn a pattern 90°, then 180°, then 270°. Ask: *What do you notice? What is the same? What is different?*

Copy and continue my pattern

- Provide pattern tracks from PCM 24, and coloured pencils, crayons or felt-tip pens. Invite children to play in pairs. Each child draws an AB or ABC repeating pattern on a track. When they have drawn at least three units of repeat, ask children to take a new track each, and copy and continue each other's pattern.

Sound and action ABC patterns (adult-led)

- Repeat the last part of the *Teaching* activity with smaller groups of children (ask children to copy, continue and create sound and action AB and ABC patterns). This could be done outdoors.

Maths Foundation Activity Book B (adult-led)

Page 15 – Patterns

When children have completed the page, help them to rotate it 90°, then 180°, then 270°. Ask: *What do you notice? What is the same? What is different?*

Session 5: ABB patterns using objects

Children observe, copy, continue and create ABB repeating patterns made from objects.

You will need:

selection of objects suitable for pattern-making, e.g. beads and laces, 2D shapes (circles, triangles, squares, rectangles), pegboards and pegs, interlocking cubes, counters, building blocks, playdough and cutters, natural materials (leaves, sticks, stones, pine cones), small-world resources in different colours and sizes

Getting started

- Before the session, place some objects suitable for pattern-making on a table. Include a range of objects that can be used to create ABB patterns in different contexts, e.g. shape, colour, size, type.
- Say the action rhyme *Clap your hands* (27). At the end of the rhyme, perform an ABB pattern of claps, e.g. clap your hands together once, then 'clap' them twice on your knees (repeated three times). Repeat several times, with children joining in.

Teaching

- Choose some objects from the table, such as big and small plastic bears. Say: *Watch carefully, and tell me what you notice.* Make an ABB pattern using a small and two big bears. Name the objects as you place each one in the row: *Small bear, big bear, big bear.* Show at least three units of repeat. Ask: *What do you notice?*
- Discuss responses. Say: *This is a repeating pattern of two different types of bear: small bears and big bears.*
- Show children the unit of repeat. Indicate each 'unit' (a small bear and two big bears) with your hands. Say: *This is a small-big-big pattern. These three bears keep repeating. See: here is another small-big-big pattern.*
- Repeat above, making a different ABB pattern using two different types/colours/sizes of object. Make sure to show at least three units of repeat.
- Repeat a third time. Then ask: *If I wanted to continue this pattern, what type of object would I put down next? What would be the*

Teaching notes

next object after that? *And the next?* Continue until the pattern has at least 15 objects.

- Repeat above, but this time make an error, e.g. red bear, blue bear, blue bear, red bear, blue bear, blue bear, red bear, red bear, blue bear. Ask: *Is this right? What have I done wrong? What colour should this bear be?*

Explore in groups

Pattern-making (using objects)

- Repeat the *Pattern-marking (using objects)* activity from Session 1 (page 106), but ask children to create ABB patterns. See Session 1 'You will need' list for suggested resources.

Copy my pattern

- Provide counters. Invite children to play in pairs. Each child creates an ABB repeating pattern of counters in two colours. When they have created at least three units of repeat, ask children to copy each other's pattern.

Continue my pattern

- Provide pegboards and pegs. Invite children to play in pairs. Each child creates an ABB repeating pattern on their board using pegs in two colours. (They could continue their pattern around the outer edges of the board.) When they have created at least three units of repeat, ask children to swap boards and continue each other's pattern. If needed, prompt them to continue for at least two units of repeat.

Session 6: AB, ABC and ABB patterns involving objects, pictures, sounds and actions

Children observe, copy, continue and create AB, ABC and ABB repeating patterns. They use objects, pictures, sounds and actions.

You will need:

Maths Foundation Reading Anthology; Beads and laces Digital Tool; selection of resources suitable for pattern-printing, e.g. stickers, self-inking stamps, plastic 3D shapes (cylinder, pyramid, cube, cuboid), stamp pads; coloured pencils, crayons and felt-tip pens; paint, water, aprons; PCM 24; chime bars/xylophone; *Maths Foundation Activity Book B*

Getting started

- Before the session, place a selection of resources suitable for pattern-printing on a table. Include a range of resources that can be used to create patterns based on shape, colour, size and type.
- Show pages 14 and 15 (Carpet patterns) of the *Reading Anthology*. Briefly remind children of the three different horizontal AB patterns, and the two different vertical ABC patterns on page 14. Then focus on the two different vertical ABB patterns on page 15. Ask children to identify and describe each pattern, and say how it would continue. Point out the unit of repeat in each pattern.

Teaching

- Display the Beads and laces Digital Tool. Show 1 lace. Use red and yellow beads to make an ABB pattern. Show three units of repeat.
- Ask children to identify the pattern (e.g. *It goes red, yellow, yellow, red, yellow, yellow*). Point out the unit of repeat, e.g. *The part that repeats is one red bead and two yellow beads. What bead comes next?*
- With children's help, continue the pattern for at least two units of repeat.
- Click 'Clear all'. Ask a child to come and choose two types of bead to start an ABB pattern, e.g. blue, blue, yellow. Say: *This is the part of the pattern that will repeat.* Add two more units of repeat. Ask another child to come and continue the pattern.
- Demonstrate a sound or action ABB pattern, e.g. clap, stomp, stomp, clap, stomp, stomp, clap, stomp, stomp. Say the pattern aloud and encourage children to join in. Point out the unit of repeat.
- Repeat for other sound and action ABB patterns, e.g. tap shoulders, tap head, tap head; step, jump, jump; tap knees, clap, clap; music patterns (using chime bars/xylophone); numbers (e.g. *one, two, two*); phonic sounds (e.g. *sh, ch, ch*).
- Ask a child to make their own sound or action ABB pattern. Make sure they include at least three units of repeat. Ask other children to copy and continue the pattern.
- Repeat with one or two more children.

Explore in groups

Pattern-making (using objects)

- Repeat the *Pattern-marking (using objects)* activity from Session 1 (page 106), but ask children to create AB, ABC or ABB patterns. See Session 1 'You will need' list for suggested resources.

Pattern-making (using pictures)

- Provide pattern tracks from PCM 24 and a range of resources that children can use to create patterns by printing (e.g. stickers, self-inking stamps, plastic 3D shapes (cylinder, pyramid cube, cuboid), paint, stamp pads). Invite children to create AB, ABC or ABB repeating patterns. Ask them to tell you about each pattern. Help them to identify the unit of repeat. Turn a pattern 90°, then 180°, then 270°. Ask: *What do you notice? What is the same? What is different?*

Copy and continue my pattern

- Provide pattern tracks from PCM 24, and coloured pencils, crayons or felt-tip pens. Invite children to play in pairs. Each child draws an AB, ABC or ABB repeating pattern on a track. When they have drawn at least three units of repeat, ask children to take a new track each, and copy and continue each other's pattern.

Sound and action ABB patterns (adult-led)

- Repeat the last part of the *Teaching* activity with smaller groups of children (ask children to copy, continue and create sound and action AB, ABC and ABB patterns). This could be done outdoors.

Maths Foundation Activity Book B (adult-led)

Page 16 – Patterns

When children have completed the page, help them to rotate it 90°, then 180°, then 270°. Ask: *What do you notice? What is the same? What is different?*

Sessions 7 and 8: Sort, represent and describe objects

Children sort, represent and describe a range of different objects using criteria such as shape, size, colour and type.

It is recommended that the same session is taught on two consecutive days. Vary the sessions slightly by using different objects and criteria for sorting. Page 17 from the *Maths Foundation Activity Book B* should be completed in Session 8.

You will need:

Maths Foundation Reading Anthology; sorting hoops; sorting trays; selection of objects suitable for sorting, e.g. buttons, beads, 2D shapes (circles, triangles, squares, rectangles), interlocking cubes, counters, building blocks, natural materials (leaves, sticks, stones, pine cones), Home corner props, soft toys, small-world resources in different colours and sizes; *Maths Foundation Activity Book B*; paper; coloured pencils or crayons

Getting started

- Before the session, place a selection of objects suitable for sorting on a table. Include a range of different objects that can be sorted according to different criteria such as shape, colour, size and type.
- Show pages 30 and 31 (Selvan's bakery) of the *Reading Anthology*. Discuss the picture with the children. Focus on how the breads and cakes/pastries have all been sorted into sets according to type. Discuss the similarities and differences between each set. Ask children to say what is the same and what is different. Emphasise the criteria by which the breads/cakes have been sorted, e.g. *All of these cakes have pink icing. These are all chocolate cakes.*

Teaching

- Ask several children to come to the front. Sort them into two sets. Use criteria such as boys/girls; glasses/no glasses; dark hair/not dark hair; jumper/no jumper. Ask: *What can you tell me about all the children in this set? What about all the children in this set? Why are these children in this set? What is the same about all the children in this set? Why is Holly in this set and not in that set?*

Teaching notes

- Then ask children to suggest different ways you could sort the same group of children.
- Then sort the children into three or four sets. Discuss the similarities and differences between each set, emphasising the criteria by which the children have been sorted.
- Ask children to suggest other ways to sort the same group of children into three or four groups.
- Next, choose a group of objects from the table. Sort them using criteria such as shape, colour, size or type. Ask: *What is the same about all the objects in this set? How are they different to the other sets? How else could we sort these objects?*
- Repeat for different groups of objects. Make sure to:

  - ask questions that help children consider what is the same about all the objects in one set, and how they are different to the other sets

  - emphasise the criteria by which the objects have been sorted

  - sort groups of objects into two, three or more sets

  - sort the same group of objects in different ways

  - encourage children to come up with their own criteria for sorting different groups of objects into sets.

Explore in groups

Sorting objects

- Provide sorting hoops, sorting trays, and a range of objects suitable for sorting, e.g. buttons, beads, 2D shapes, interlocking cubes, counters, building blocks, natural materials, Home corner props, soft toys, small-world resources in different colours and sizes. Invite children to sort objects using their own criteria. Encourage them to think about shape, colour, size and type, and to consider sorting

into two, three or more sets. Ask questions such as: *What is the same about all the objects in this set? How are the objects in this set different to the ones in the other sets? Could you sort these objects in a different way?*

Represent a sorted group of objects

- Provide sorting trays, a range of small objects suitable for sorting, paper and pencils. Invite children to take some objects and sort them using their own criteria. Then encourage children to represent their sorted objects on paper.

Maths Foundation Reading Anthology (adult-led)

- Together, look at pages 30 and 31 (Selvan's bakery) of the *Reading Anthology*. Repeat the *Getting started* activity with smaller groups of children.

Maths Foundation Activity Book B (adult-led)

Page 17 – Sort

When children have completed the page, ask them to suggest other ways that the buttons could be sorted.

Assessment opportunities

Assess children's learning against the objectives for this unit, using the guidance on formative assessment on pages 24–25, and record your observations in the Unit 9 progress tracking grid on page 30. The relevant pages of *Activity Book B* can also be used for assessment.

Can the children:

- observe, copy, continue and create AB, ABC and ABB patterns made with objects, pictures, sounds and actions?
- spot an error in a pattern and correct it?
- begin to identify the unit of repeat in a pattern?
- sort, represent and describe a range of different objects using criteria such as shape, colour, size and type?

Unit 10 Time

Theme: Day and night

Overview

This unit introduces children to the concept of time. They will use positional language such as now, before, soon, next, after, later, last, early and late. They will also use relative terms such as morning, afternoon, evening, night, yesterday, today and tomorrow.

Children begin to sequence events familiar to them. This will include events in well-known stories, and important events and actions in family life and school routines.

Children are also introduced to the days of the week and begin to establish awareness of the correct sequence and the cycle of the days of the week.

Learning objectives

Geometry and Measure – Time	• **9a Begin to understand and use the vocabulary of time, including the days of the week, yesterday, today, tomorrow, morning and evening.** • **9b Sequence familiar events.**

Learning objectives in bold are taught for the first time in this unit.

Vocabulary

time, early, earlier, before, now, soon, next, after, late, later, last, bedtime, dinnertime, playtime, morning, afternoon, evening, night, tonight, days, week, Monday, Tuesday, Wednesday, Thursday, Friday, Saturday, Sunday, first, order, today, yesterday, tomorrow

Making connections

English: Day and night

Science: Light and dark

Preparation

You will need:

- *Maths Foundation Reading Anthology*, pages 26–29.
- *Maths Foundation Activity Book B*, pages 18–23.
- Unit 10 slides.
- PCMs 28–31.
- Rhymes and songs: *Seven days in the week* (28), *Sing with me the days of the week* (29) (page 219).
- Familiar story that has a clear sequence of events.
- The story of *The Enormous Turnip* (also called *The Gigantic Turnip* or *The Big Turnip*) (or adapt Session 2 if there is a story that children are more familiar with that has a clear sequence of events).
- Suitable props that feature in the above stories.
- Large sheets of paper.
- Magazines, newspapers, catalogues, etc.
- Scissors.
- Glue.
- Easel or board and magnets or sticky tack.
- Painting and drawing station: paper, pencils, coloured pencils and crayons, paint, water, paintbrushes, aprons.

Teaching notes

Before starting

Children will get the most out of this unit if they already have a developing awareness of time. For example, they recognise regular events in the school day, such as snack or lunchtime.

It is recommended that *Teaching* from Session 6 is repeated more than once during this unit.

The teaching of time should be ongoing, and not limited to a single unit. It is recommended that the *Teaching* sessions and *Explore in groups* activities in this unit are adapted and repeated, at different times throughout the year.

Common difficulties

Time is an abstract concept. It can not be measured in a physical or visual way (unlike length, mass and capacity). Therefore children often find it difficult. Young children also live more fully in the present, focused on the 'here and now'. Encourage children to remember and imagine events in the past and future that are based on familiar events that are personal to them.

Session 1: The concept of time

Children begin to understand the concept of time. They describe familiar events using words such as 'early', 'earlier', 'before', 'now', 'soon', 'next', 'after', 'later' and 'last'.

You will need:

Unit 10 Session 1 slides; familiar story that has a clear sequence of events; suitable props that feature in the story (optional); PCM 28 (ideally enlarged to A3); *Maths Foundation Reading Anthology*; paper; coloured pencils and crayons; paint, water, paintbrushes, aprons

Getting started

- Begin by holding a discussion with the children about their day. Encourage them to talk about experiences personal to them. Discuss past, present and future events in the day, asking questions such as: *What did you do early this morning? Tell me something we all did earlier. What did we do just before we came and sat on the floor? Tell me something we might do later today. What is the last thing you do at night?* Emphasise the mathematical language associated with time.

Teaching

- Display slide 1. Explain to children that these slides tell the story of Asha's day. Slowly go through slides 1 to 8. Discuss what is happening in each scene. Highlight the following events.

  - Slide 1: Asha is at home having breakfast.

  - Slide 2: Asha is arriving at school.

  - Slide 3: Asha is in her classroom.

  - Slide 4: Asha is playing with some of her classmates.

  - Slide 5: Asha is leaving school.

  - Slide 6: Asha is at home, after school, playing in the garden.

  - Slide 7: Asha is having dinner.

  - Slide 8: Asha is in bed asleep.

- Then show slide 9. Discuss and ask questions related to the sequence of events in Asha's day. Use mathematical language associated with time (e.g. early, earlier, before, next, after, later, last).
- Briefly discuss with children the similarities and differences between Asha's day and their day (at school).
- Discuss a familiar story that has a clear sequence of events. If appropriate, hold up the book or a suitable prop to remind children of the story. Ask questions such as: *What happened before/after …? What happens next/last?*

Explore in groups

Asha's day (adult-led)

- Shuffle the cards from PCM 28 (Asha's day). Show the cards one at a time to the children.

Ask them to tell you what is happening in each picture. Then spread out all the cards face up. Point to the first event (Asha having breakfast), and ask children to tell you the next thing that Asha did. Continue, placing the cards in chronological order. Use the words: 'next', 'after', 'later' and 'last'. Once all eight cards are in order, point to different cards and ask questions using the words: 'early', 'earlier', 'before', 'next', 'after', 'later' and 'last'.

Painting and drawing station

- Invite children to paint or draw a picture of something that they did earlier in the day (or something they are likely to do later).

Maths Foundation Reading Anthology
(adult-led)

- Together, look at pages 26 and 27 (Lee's day) of the *Reading Anthology*. Discuss and ask questions about Lee's day, using mathematical language associated with time (e.g. early, earlier, before, next, after, later, last).

Session 2: Identify and sequence familiar events

Children begin to sequence familiar events, including those in well-known stories.

Before starting

This session focuses on the story of *The Enormous Turnip* (also called *The Gigantic Turnip* or *The Big Turnip*). However, this session should be adapted if children are more familiar with a different story that has a clear sequence of events.

You will need:

Unit 10 Session 2 slides (optional); story of *The Enormous Turnip* (or a similar story); suitable props that feature in the story (optional); paper; coloured pencils and crayons; PCMs 28 and 29

Getting started

- If possible, have a book of the story to read to the children and/or some suitable props to show. Alternatively, use the Unit 10 Session 2 slides.
- Read/tell the story of *The Enormous Turnip*.

Teaching

- Discuss the story with the children: *What happens at the beginning of the story? Then what happens? What happens next? What happens at the end?*
- Highlight the sequence of events:

  1. A farmer planted a turnip. The turnip grew and grew. It grew to be an enormous turnip. The farmer tried to pull the turnip out of the ground. He pulled and pulled, but couldn't pull it out.

  2. So, he called his wife to help. They pulled and pulled, but couldn't pull it out.

  3. Then, the wife called over their grandchildren. They all pulled and pulled, but couldn't pull it out.

  4. Next, a granddaughter called over the dog and the cat. They all pulled and pulled, but couldn't pull it out.

  5. After that, the cat called over the mouse. They all pulled and pulled and pulled.

  6. Finally, out came the enormous turnip!

Explore in groups

The Enormous Turnip

- Provide a set of shuffled cards from PCM 29 (The Enormous Turnip). Invite children to work together to place the cards in chronological order. Ask them to explain what is happening and to describe the sequence of events. Then ask the children to close their eyes. Change the order of two of the cards. Ask children to tell you what they notice. Are they able to put the cards back into the correct order?

Variation: use cards from PCM 28 (Asha's day) instead.

Our school day (adult-led)

- Discuss regular, significant events in the school day, e.g. the start of school, show and tell, playtime. Then ask individual children to tell you their favourite event of the day. (Alternatively, tell each child an event.) Ask each child to draw the event. Once all the children in the group have done this, bring the children back together. Ask the children to help you place the drawings in chronological order.

Teaching notes

Three-event sequence (adult-led)

- Discuss a three-event sequence, such as growing a flower (1. planting the seed; 2. stem and leaves grow; 3. flower blooms) or making an ice cream (1. get the cone; 2. add a scoop of ice cream; 3. add toppings). After the discussion, ask each child to draw the three stages (e.g. empty plant pot; pot with stem and leaves; pot with stem, leaves and flower).

Variation: ask each child to draw only one part of the three-event sequence.

Session 3: Times in a day

Children begin to recognise important times in their day such as bedtime, dinnertime and playtime.

You will need:

Maths Foundation Reading Anthology; Unit 10 Session 3 slide; PCM 28; paper; coloured pencils and crayons; paint, water, paintbrushes, aprons

Getting started

- Show pages 26 and 27 (Lee's day) of the *Reading Anthology*. Discuss the picture with the children. Ask them to describe what they can see in the picture. Focus on mathematical language associated with time (e.g. early, earlier, before, now, soon, next, after, later, last).

Teaching

- Discuss the various important 'times' in Lee's day: time to get up, time to get dressed, time for school, bath time, bedtime. Discuss and ask questions related to the sequence of events.
- Ask individual children to talk about significant times in their own day. Ask them to say which of the things that they do are the same as, and different to, Lee.
- Display slide 1. Repeat above for Asha's day.

Explore in groups

Maths Foundation Reading Anthology
(adult-led)

- Together, look at pages 26 and 27 (Lee's day) of the *Reading Anthology*. Repeat *Teaching* with smaller groups of children. Then repeat for Asha's day, using the cards from PCM 28.

Asha's day

- Shuffle a set of cards from PCM 28 (Asha's day). Spread them out face up. Invite children to work in pairs. They take turns to point to a card. Their partner describes the event using phrases such as breakfast time, time for school, playtime, home time, bedtime. Children continue until they have described all eight cards.

Painting and drawing station

- Invite children to paint or draw a picture of themselves at bedtime, dinnertime, playtime or another time of the day. Ask children to tell you what they are drawing/painting and when during the day this happens. When children have finished, write the time on their pictures (e.g. 'bedtime', 'dinnertime', 'playtime').

Session 4: Stages of a day

Children identify familiar events that occur at different stages of a day: morning, afternoon, evening and night.

You will need:

Unit 10 Session 4 slide; *Maths Foundation Reading Anthology*; paper; coloured pencils and crayons; magazines, newspapers, catalogues; large sheets of paper; scissors; glue; *Maths Foundation Activity Book B*

Getting started

- Introduce children to the different stages of a day: morning, afternoon, evening and night. Discuss each stage in turn, talking about things that you do during that stage of the day. During this discussion, refer to the sun and moon, and use words and phrases such as 'light', 'bright', 'dark', 'daytime' and 'during the day'.
- Then ask individual children to talk about the things that they do during the different stages of a day.

Teaching

- Display slide 1. Discuss each event in Asha's day. Ask children to identify at what stage of a day each event happens (morning, afternoon, evening or night).
- Show pages 26 and 27 (Lee's day) of the *Reading Anthology*. Repeat above for Lee's day.

Explore in groups

Daytime and night-time posters

- Provide large sheets of paper, scissors, glue and a selection of magazines, newspapers, catalogues, etc. Invite children to select a variety of daytime and night-time pictures (of scenes, or objects associated with daytime and night-time). Ask them to make two different posters: a daytime poster and a night-time poster. (Alternatively, they could make just one poster: either daytime or night-time.) Ask children to tell you about their poster.

Asha's day

- Provide a shuffled set of cards from PCM 28 (Asha's day). Invite children to sort the cards into four groups: morning, afternoon, evening and night. Ask children to tell you how they decided which set each card belongs in.

Variation: ask children to sort the cards into two groups: daytime and night-time.

When? (adult-led)

- Act out an activity that you would do either in the morning, afternoon, evening or night, e.g. sitting up, yawning and stretching to indicate waking up. Ask children to guess what you were acting out, and say when in the day it would happen. Demonstrate one or two more examples. Then ask children to take turns to think of an activity and act it out. When appropriate, discuss the fact that some events happen during more than one stage of a day, e.g. brushing your teeth or eating a meal.

Maths Foundation Activity Book B (adult-led)

Page 18 – Morning and night

Session 5: Days of the week

Children begin to learn the days of the week.

You will need:

Unit 10 Session 5 slides; *Maths Foundation Reading Anthology*; paper; coloured pencils and crayons; paint, water, paintbrushes, aprons

Getting started

- Display the slides (choose from slide 1, 2 or 3 depending on which day of the week is considered the first day).
- Teach children one of the following rhymes: *Seven days in the week* (28) or *Sing with me the days of the week* (29). Say the rhyme several times before asking children to join in.
- Please note that it is not expected at this stage that children read the days of the week. However, displaying the words will help to familiarise children with the names and sequence.

Teaching

- Keep the slide on display. Talk about the different things that you do on certain days of the week. Tell children about things that you only do on certain days, e.g. *I go shopping every Thursday. On Tuesdays, I visit my parents. I always go swimming on Wednesday.* Also talk about things that happen on more than one day each week, (e.g. coming to school), and say on which days they happen. As you say a particular day of the week, point to it on the slide.
- Then ask individual children to talk about the things that they do on different days of the week. Prompt and assist them where needed. As the child talks about an event, point to the day of the week on the slide.
- Show pages 28 and 29 (Luke's exercise diary) of the *Reading Anthology*. Discuss the pictures with the children. Draw attention to the specific exercise that Luke does each day. Then ask questions such as: *What exercise does Luke do on Friday? On which day of the week does Luke ride his bike? What does he do the day after/before he goes rollerblading?*

Explore in groups

Painting and drawing station

- Invite children to draw or paint a picture of something that they do on a specific day of the week. This may be something they do at school, or at home. Before they start to draw/paint, ask children to tell you what event they have chosen. If they are able, ask them to tell you on which day this happens. If children are not able to identify the day, lead them towards remembering a certain event that happens on

Teaching notes

a particular day. When children have finished, write the day of the week on their pictures.

Maths Foundation Reading Anthology
(adult-led)

- Together, look at pages 28 and 29 (Luke's exercise diary) of the *Reading Anthology*. Repeat the last part of the *Teaching* activity with smaller groups of children.

Maths Foundation Reading Anthology

- Give children a copy of the *Reading Anthology* opened at pages 28 and 29 (Luke's exercise diary). Tell children what day it is today. Ask them to draw a picture of themselves doing the same exercise that Luke does on that day. When children have finished, write the day of the week on their pictures.

Variation: assign each child a different day of the week.

Session 6: The cycle of days in the week

Children begin to learn about the cycle of days in the week.

You will need:

Unit 10 Session 6 slides; PCM 30; large sheets of paper; glue; *Maths Foundation Activity Book B*; coloured pencils

Getting started

- Prepare enough copies of the cards from PCM 30 (days of the week) to be able to give one card to each child.
- Remember, it is not expected at this stage that children read the days of the week. However, displaying the words will help to familiarise children with the names and sequence.
- Display the slides (choose from slide 1, 2 or 3 depending on which day of the week is considered the first day).
- Say one of the following rhymes with the children: *Seven days in the week* (28) or *Sing with me the days of the week* (29).

Teaching

- Keep the slide on display. Give each child a day of the week card.
- Ask a child whose card shows the first day of the week (Saturday/Sunday/Monday, as

relevant) to come and stand at the front. Ask: *Who can tell me what day of the week Leo has?* Prompt children if necessary. *Put up your hand if your card shows the same day as Leo's. Well done, you all have …day.*

- Then ask a child who has a card showing the next day of the week to come to the front. Ask them to stand to the right of the first child (as viewed by the rest of the children). Ask: *Who can tell me what day of the week Ayesha has? Put up your hand if your card shows the same day as Ayesha's. That's right, you all have …day.*
- Point to the two children at the front, and say: *…day is the day after …day. And …day is the day before …day.*
- Repeat for the remaining five days of the week.
- Then go and stand behind a child. Ask: *What day does Chaiya have?* (e.g. *Friday*) *What day comes after Friday? What day is before Friday?* Repeat several times.
- Then along with all children, chant the days of the week, starting with the first day of the week. Repeat.
- Stand behind the child holding the first day of the week card. Say the day's name. Ask all the children whose card shows the same day of the week to stand up and pass you their card. Continue until you have collected all of the cards.

Explore in groups

Order the days of the week cards

- Provide shuffled sets of cards from PCM 30 (days of the week). Invite children to take a set of cards and work in pairs. They take turns to place the cards in order, starting with the first day of the week. The other child checks their partner's order. Ask them to repeat so that each child orders the cards twice. Prompt the children to shuffle the cards after each turn.

Days of the week poster

- Provide large sheets of paper, glue and shuffled sets of cards from PCM 30 (days of the week). Invite children to make a poster in small groups. They take a set of cards and stick them in order on a large sheet of paper. They could position them in a circle, similar to the cycle on the slides.

Maths Foundation Activity Book B (adult-led)

Page 19 – Days of the week

Session 7: Describe when events happen

Children begin to describe when events happen using words such as 'today', 'yesterday' and 'tomorrow'.

You will need:

Unit 10 Session 7 slides; PCM 30; *Maths Foundation Reading Anthology*; *Maths Foundation Activity Book B*; coloured pencils

Getting started

- Display the slides (choose from slide 1, 2 or 3 depending on which day of the week is considered the first day).
- Say one of the following rhymes with the children: *Seven days in the week* (28) or *Sing with me the days of the week* (29).

Teaching

- Keep the slide on display. Ask: *Does anyone know what day of the week it is today?* Emphasise the word 'today'. Take responses. If needed, help children to identify the correct day.
- Circle (or draw a star beside) the correct day on the slide, e.g. Wednesday. Say: *Today is Wednesday.*
- Discuss with children some of the things that they have already done today.
- Point to the day on the slide. Say: *Today is Wednesday. The day before today was Tuesday.* Emphasise the word 'before'. Point to the day before on the slide. Say: *We can also say that yesterday was Tuesday.* Emphasise the word 'yesterday'.
- Discuss with children some of the things that they did yesterday at school (if children were in school). Then ask them to tell you some of the things that they did yesterday at home.
- Refer back to today's day of the week: *So, today is Wednesday. The day after today will be Thursday.* Emphasise the word 'after'. Point to the day after on the slide. Say: *We can also say that tomorrow will be Thursday.* Emphasise the word 'tomorrow'.
- Discuss with children some of the things that they are likely to do tomorrow at school (if children will be at school). Then ask them to tell you some of the things that they are likely to do tomorrow at home.
- Show pages 28 and 29 (Luke's exercise diary) of the *Reading Anthology*. Ask children to tell you what exercise Luke will be doing today, what exercise he did yesterday and what exercise he will do tomorrow.

Explore in groups

Yesterday or tomorrow (adult-led)

- Shuffle a set of cards from PCM 30 (days of the week). Place them face down in a pile. Turn over the top card (e.g. Tuesday). Say: *This card says Tuesday. If today is Tuesday, what day was it yesterday?* Choose another card (e.g. Friday). Say: *If today is Friday, what day will it be tomorrow?* Repeat several times, reshuffling the cards when necessary.

Yesterday and tomorrow

- Ask children to repeat the *Yesterday or tomorrow* activity above, but working in pairs. It is recommended that children take part in the adult-led version first. Each pair will need their own set of cards from PCM 30. Children take turns to turn over the top card and ask their partner, e.g. *If today is Tuesday, what day was it yesterday/what day will it be tomorrow?*

Maths Foundation Reading Anthology (adult-led)

- Together, look at pages 28 and 29 (Luke's exercise diary) of the *Reading Anthology*. Discuss and ask questions about what Luke does on different days, using the words 'today', 'yesterday' and 'tomorrow'.

Maths Foundation Activity Book B (adult-led)

Pages 20 and 21 – Today, yesterday, tomorrow

Children may need help to identify which day is 'today'.

Session 8: Recall events

Children recognise the order of the days of the week. They describe events using words such as 'today', 'yesterday' and 'tomorrow'.

You will need:

Unit 10 Session 8 slides; *Maths Foundation Reading Anthology*; PCMs 30 and 31 (including an enlarged copy of each PCM); easel or board

Teaching notes

and magnets or sticky tack; *Maths Foundation Activity Book B*; pencils

Getting started

- Before the session, place the enlarged set of cards from PCM 30 (days of the week) down the right-hand side of the easel, making sure that they are not in chronological order. Place the cards from PCM 31 beside the easel.
- Arrange the children in a circle. Choose a child to start. Ask them to say the first day of the week. Then go around the circle, with each child saying the next day in the sequence. On reaching the last day of the week continue, saying the first day of the week again. If necessary, display the slides (choose from slide 1, 2 or 3 depending on which day of the week is considered the first day).

Teaching

- Briefly show pages 28 and 29 (Luke's exercise diary) of the *Reading Anthology*. Remind children of what exercise Luke does each day.
- Point to the days of the week cards on the easel. Ask: *What is the first day of the week?* Emphasise the word 'first'. Take the correct card and place it at the top of the left-side. *What day comes after …?* Take the correct card and place it directly underneath the first card.
- Continue until all seven cards are in chronological order.
- Ask: *What day is it today?* Emphasise the word 'today'. Point to the appropriate days of the week card on the easel.
- Say: *Think about the exercises that Luke does. What exercise does Luke do today?* Take responses from the children. If necessary, prompt them by suggesting two alternatives: one correct, the other incorrect. *That's right. Today is …day, and on …day Luke …* Hold up each of the cards from PCM 31 in turn until the matching exercise card is found. Place the card to the right of today's day of the week card.
- Repeat for the exercise Luke did yesterday, then the exercise he will do tomorrow. If necessary, continue to prompt children by suggesting two alternatives.
- Then move onto matching the remaining four days of the week. (The exercises for each day are: Monday swimming, Tuesday bike riding, Wednesday skateboarding, Thursday climbing, Friday trampolining, Saturday rollerblading, Sunday playing soccer/football.)

- Show pages 28 and 29 (Luke's exercise diary) of the *Reading Anthology* again. Ask children to help you check that the cards on the easel are correct.

Explore in groups

Luke's exercise diary (A)

- Provide shuffled sets of cards from PCM 31 (Luke's exercise diary). Invite children to place the cards in order, starting with the exercise that Luke did on Monday. When they have done this, give them a copy of the *Reading Anthology* opened at pages 28 and 29 (Luke's exercise diary) to check their order.

Luke's exercise diary (B)

- Provide shuffled sets of cards from PCM 30 (days of the week) and 31 (Luke's exercise diary). Invite children to match each exercise to the corresponding day of the week card. When they have done this, give them a copy of the *Reading Anthology* opened at pages 28 and 29 (Luke's exercise diary) to check that they have matched the cards correctly.

Maths Foundation Activity Book B (adult-led)

Pages 22 and 23 – Luke's exercise diary

Children may need help to read the days of the week.

Assessment opportunities

Assess children's learning against the objectives for this unit, using the guidance on formative assessment on pages 24–25, and record your observations in the Unit 10 progress tracking grid on page 31. The relevant pages of *Activity Book B* can also be used for assessment.

Can the children:

- show an awareness of time, using words such as: 'now', 'before', 'soon', 'next', 'after', 'later', 'last', 'early', 'late', 'bedtime', 'dinnertime' and 'playtime'?
- recognise and name different stages of a day: morning, afternoon, evening and night?
- identify and sequence familiar events?
- recognise and begin to order the days of the week?
- describe when events happen using words such as: 'yesterday', 'today' and 'tomorrow'?

Unit 11 Numbers to 10 (B)

Theme: Weather

Overview

This unit reinforces the five key counting principles. These were introduced in Units 1 and 2 for numbers 1 to 5, and extended in Unit 6 for numbers up to 10.

Children continue to count from 1 to 10 forwards and backwards. They count sets of objects, and things that cannot be counted. They compare quantities and numbers from 1 to 10 using the language 'more', 'less' or 'fewer'. They practise reading and writing the numerals (but not the words) from 1 to 10.

Children also begin to share objects into two equal groups in the context of play.

Learning objectives

Number – Counting and understanding numbers	• *1a Say and use the number names in order in familiar contexts such as number rhymes, songs, stories, counting games and activities, from 1 to 5, then 1 to 10.* • *1b Recite the number names in order, continuing the count forwards or backwards, from 1 to 5, then 1 to 10.* • *1c Count objects from 1 to 5, then 1 to 10.* • *1d Count in other contexts such as sounds or actions from 1 to 5, then 1 to 10.* • **1e Share objects into two equal groups.**
– Reading and writing numbers	• *2a Recognise numerals from 1 to 5, then 1 to 10.* • *2b Begin to record numbers, initially by making marks, progressing to writing numerals from 1 to 5, then 1 to 10.*
– Comparing and ordering numbers	• *3a Use language such as more, less or fewer to compare two numbers or quantities from 1 to 5, then 1 to 10.*

Learning objectives in italics have been taught previously. In this unit they are consolidated and/or extended.

Learning objectives in bold are taught for the first time in this unit.

Vocabulary

number, count, count on, count forwards, count back, count backwards, one, two, three, …, eight, nine, ten, next, after, before, how many, compare, more, less, fewer, the same, more than, less than, fewer than, equal sharing, 'fair' sharing, equal groups

Making connections

English: Weather

Science: Weather

Preparation

You will need:

- *Maths Foundation Reading Anthology* (and ebook), pages 10–13.
- *Maths Foundation Activity Book C*, pages 2–6.
- Unit 11 slides.
- Number fluency slide: Target board
- Digital Tools: Number cards, Tree.
- PCMs 1–5, 7 and 8, 11–14.
- Number fluency games and activities: *Finger counting* (1), *Continue counting* (3), *Target board (counting objects)* (11), *Marbles in a tin* (12) (pages 198–201).
- Rhymes and songs: *One potato, two potatoes* (14) (page 213).
- Large 1–10 number bunting and/or large number track.
- Circle-shaped stamps/stickers (optional).
- Trays or containers.
- Plates or dishes.
- Readily available classroom resources, e.g. washing line, pegs, socks, cards, balls, buckets, hoops.
- 10 marbles and a tin.
- Percussion instruments, e.g. chime bars, drums and sticks.
- Beanbags in four different colours.
- Whiteboard and coloured pens (optional).
- 10 empty tissue boxes (or shoe boxes or similar with a slot cut out of the top).
- Small pieces of paper.

- 1 to 6 dot dice.
- Counting collections station: a selection of different small-world resources (e.g. people, animals and sealife, transport) and counting apparatus (e.g. beads and laces, counters, interlocking cubes, pegboard and pegs), small pots or cups, trays, sorting trays, numeral cards 1–10 from PCM 5.
- Construction station: construction materials, e.g. building blocks, toy construction bricks, recycled boxes and packaging.
- Painting and drawing station: paper, pencils, coloured pencils and/or crayons.

Before starting

Display number bunting or a large number track showing numerals 1 to 10. Each numeral should have a matching picture (e.g. 9 dots below the numeral 9). Refer to the bunting/track regularly throughout the unit.

If necessary, give children copies of PCMs 7, 8, 12 and 13 to practise numeral formation and recognising cardinal values to 10.

Before teaching this unit, look back at the Assessment section at the end of Unit 6 (page 85), to identify the prerequisite learning for this unit.

Common difficulties

Children may already have some experience of sharing, e.g. sharing pencils between the tables. However, they may not yet understand *equal sharing*. Discuss the difference between 'fair' and 'unfair' sharing. For example, if we share objects into two groups, it is only 'fair' if the two groups are equal.

Maths background

The word 'fewer' is used when talking about people or things in the plural. The word 'less' is used when talking about things that are uncountable, or have no plural. It is important to use the correct mathematical vocabulary throughout this unit. When comparing two sets of objects, use the words 'more' and 'fewer'. When comparing two numbers, use 'more' and 'less'.

Session 1: Count forwards and backwards, starting from any number from 1 to 10

Children count forwards and backwards, starting from any number from 1 to 10. They say which number comes next and before.

You will need:

1 to 10 number track or number bunting; PCM 5; *Maths Foundation Activity Book C*; pencil

Getting started

- Practise counting forwards and backwards from 1 to 10, using one of two Number fluency activities. Depending on the ability of the children, use either *Finger counting* (1) or *Continue counting* (3).
- Ensure that all children can clearly see the 1 to 10 number track/bunting.

Teaching

- Point to 3 on the number track/bunting. Say: *We're going to count to 10, but this time we're not going to start from 1. We're going to count forwards from 3. Ready? Let's go!*
- Repeat several times, counting forwards to 10 from a number other than 1.
- Without referring to the number track, ask questions such as: *What number comes after 4/9/3 …? Tell me the number that comes before 5/8/3 …? What's the next number in this count: 3, 4, 5 …? Tell me the number that comes next: 6, 7, 8 …*
- Then move on to counting backwards. Rehearse counting back from 10 to 1.
- Then say: *Now we're going to count back to 1 again, but we're not going to start from 10.* Point to 7 on the number track/bunting: *We're going to count backwards from 7. Ready? Let's go!*
- Repeat several times, counting backwards to 1 from a number other than 10.
- Ask questions such as: *What's the next number in this count: 7, 6, 5, …? What number comes next: 9, 8, 7 …?*

Explore in groups

Repeat Teaching (adult-led)

- Repeat the *Teaching* activity with smaller groups of children. Alter the activity slightly by

using a shuffled set of numeral cards 1–9 from PCM 5. Place the cards face down in a pile on the table. Turn over the top card. Ask children to count forwards to 10, or backwards to 1, from the number.

Which number comes next/before?

- Provide sets of shuffled numeral cards 1–9 from PCM 5. Invite children to play a game in pairs. They spread a set of cards face down on the table. One child turns over a card. The other child says the number that comes *next* when counting forwards. Continue until all nine cards have been used.

Variation: use numeral cards 2–10 from PCM 5, children say the number that comes *before*.

Maths Foundation Activity Book C (adult-led)

Pages 2 and 3 – Trace and join

When children have completed both pages, ask them to count backwards from 10 to 1. Then ask them to count forwards to 10 from a number other than 1. Finally ask them to count backwards to 1 from a number other than 10.

Session 2: Match a numeral to a set of up to 10 objects

Children continue to use the numerals 1 to 10 to represent quantities.

You will need:

Maths Foundation Reading Anthology; Number cards Digital Tool; PCMs 1–5, 8, 11 and 13; coloured pencils or (if available) circle-shaped stamps/stickers

Getting started

- Show pages 10 and 11 (How many animals?) of the *Reading Anthology*. Discuss the pictures with the children. Point to the different animals in the picture. Ask children to tell you how many there are.

Teaching

- Shuffle all the cards from PCMs 1–4. Give one card to each child. Ask children to count the objects on their card.
- Display the Number cards Digital Tool. Randomly arrange cards 1–10 on the board

Teaching notes

and hide all the values. Say: *I'm going to show you a number on one of these cards. If you have counted the same number of objects on your card, hold it up.*

- Click on one of the number cards, e.g. 3. All the children with three objects on their cards hold them up. Ask: *If you are holding up a card, what's my number?*
- Repeat with the remaining nine number cards.
- If time allows, ask the children to pass their cards around the class/group until you say stop. Hide all the values of the cards on the Number cards Digital Tool and swap the positions of the cards. Repeat the activity.

Explore in groups

Practising numeral formation and recognising cardinal values

- Before the activity, prepare copies of PCM 13. On each sheet, circle three numbers. Use a different coloured pencil for each circle, e.g. a blue circle around 6, a red circle around 8 and a green circle around 10. Give each child one of these sheets and three coloured pencils to match the colours of the circles. Ask children to trace each circled number in the corresponding colour. Then ask them to colour that number of stars in the same colour (e.g. they colour 6 stars blue, 8 stars red and 10 stars green).

Variation: use PCM 8 instead. Children practise writing the numbers 1 to 5 and colour balloons to match.

Making ten-frame cards (adult-led)

- Before the activity, prepare a set of blank ten-frame cards from PCM 11. On the back of each card, write a number from 1 to 10. Ensure that there are about three cards for each child. Ask children to look at the number on the card and then represent the number on the ten-frame on the other side. They could either draw large dots or (if available) use circle-shaped stamps/stickers.

Pairs

- Provide sets of cards from PCMs 1 (objects) and 5 (numerals). Encourage children to play Pairs (see Generic games rules on page 220). Each pair or group will need six objects cards and the six corresponding numeral cards.

Variation: use cards from PCM 2, 3 or 4 instead of PCM 1.

Session 3: Count out a specified number of objects up to 10 from a larger group

Children count out a specified number of objects from a larger group, not just counting the number of objects in the group.

You will need:

tray or container; selection of different small-world resources; plate or dish; PCM 5; paper; coloured pencils; readily available classroom resources, such as beads and laces, counters, sorting trays (or egg cartons), pegboard and pegs, washing line, pegs, socks, cards; construction materials such as building blocks

Getting started

- Before the activity, place these items on a table: a tray of around 20 small-world people, a plate and numeral cards 1–10 from PCM 5 (optional).
- Say the action rhyme *One potato, two potatoes* (14) to practise counting forwards from 1 to 10.

Teaching

- Before the activity, place these items on a table: a tray of around 20 small-world people, a plate and numeral cards 1–10 from PCM 5 (optional).
- Say a number from 1 to 10 (or show a numeral card), e.g. 7.
- Ask a child to come and count out that number of small-world people from the tray. As they count each one, they move it onto the plate.
- Then ask the rest of the children to count each small-world person as you pick it up from the plate and return it to the tray. Confirm the count: *Gopal counted out 7 people.*
- Say: *I am going to count to 7. When I get to '7', I want you to hold up 7 fingers and shout out the number 7.*
- Repeat several times. Say (or show) a different number each time.

Explore in groups

Counting collections station

- Provide small-world resources, organised into collections (with more than 10 resources in each collection). Place a numeral card 1–10 from PCM 5 next to each collection. Invite children to choose a collection, and draw a suitable picture, e.g. a house for people; a field for farm animals; a road for vehicles; a fish tank for sea creatures. Then ask children to look at the numeral card next to their collection. Invite them to count out that number of resources from the collection, and place them on their drawing.

Counting collections station (adult-led)

- Provide counting apparatus and other classroom resources, organised into collections (with more than 10 resources in each collection). Place a numeral card 1–10 from PCM 5 next to each collection. Invite children to choose a collection and look at the numeral card to see how many they need to use. For example, ask children to: thread 9 beads onto a lace; put 5 counters into the sorting tray; make a row of 4 pegs on the pegboard; peg 7 socks (or cards) onto a washing line.

Construction station

- Tell each child the number 7, 8 or 9. Invite children to build a model using that number of pieces. They can use whatever construction materials are available. Compare the completed models. Ask: *Which models all have 7 pieces?* Repeat for models with 8 and 9 pieces.

Session 4: Count sounds and actions to 10

Children continue to count non-physical things such as sounds, actions and remembered or imaginary objects: *abstraction*.

You will need:

10 marbles and a tin; PCM 5; percussion instruments; beanbags in four different colours; paper; whiteboard and coloured pens; paper, pencils, coloured pencils

Getting started

- Use the Number fluency activity: *Marbles in a tin* (12). Slowly drop 8 marbles into the tin. When finished, tell children that they need to try and remember that number (for one of the *Explore in groups* activities).

Teaching

- Choose a child to come to the front. Give them a percussion instrument, e.g. a chime bar.
- Show the child a numeral card from PCM 5, e.g. 6. (If necessary, also whisper the number to them.) Say: *I want you to slowly and clearly play that number of beats on the chime bar. I want the rest of you to close your eyes and listen very carefully. Count the beats that you hear. But don't count out loud; I want you to count in your heads.*
- Once the child has played the correct number of beats, ask the rest of the children to open their eyes. Ask them to hold up the corresponding number of fingers, e.g. 6.
- Then ask the child with the instrument: *Neema, how many beats did you play on the chime bar?* (6) Confirm that the child is right by holding up the numeral 6 card.
- Say to the rest of the children: *Neema made 6 beats with the chime bar and you are all showing me 6 fingers.*
- Repeat with a different child. This time, ask the child to do an action, e.g. take 9 steps.
- Repeat the activity several more times. Choose a different child, number and sound/ action each time.

Explore in groups

Counting bean bags (adult-led)

- Before the activity, place a selection of beanbags in four different colours around the room (or outdoors). Ensure that all the beanbags are visible. Ask children to secretly count all the red beanbags they can see (do not allow them to touch the beanbags). Ask children to tell you how many they counted. On paper (or a small whiteboard) draw a red beanbag and write the number(s) offered by the children. Repeat for the other three beanbag colours. Then send some children off to collect all the red beanbags. Together, count these. Compare the actual number to the numbers written on the paper/whiteboard. Repeat for the other three beanbag colours.

Teaching notes

Painting and drawing station

- Ask: *Can you remember how many marbles I dropped into the tin earlier?* Invite children to draw a picture of the tin containing the (8) marbles.

Music makers

- Place the numeral cards (1–10) from PCM 5 in a pile face down on the table. Provide percussion instruments. Encourage children to play a game in pairs or as a group. Children take turns to look at the top card. They do not show the card to the other children. The child makes the corresponding number of sounds using a percussion instrument. When they have finished, the other children say the number of beats they heard. All the children look at the numeral card to check.

Session 5: Select the correct numeral to represent up to 10 objects

Children select the correct numeral to represent a group of 1 to 10 objects.

You will need:

Target board slide; Unit 11 Session 5 slides; PCMs 1–5; small-world resources; readily available classroom resources such as beads and laces, counters, interlocking cubes, pegboard and pegs, pots, bowls; 10 empty tissue boxes (or shoe boxes or similar with a slot cut out of the top)

Getting started

- Use the Number fluency activity *Target board (counting objects)* (11). Display the Target board slide. Focus on counting the objects. Ask questions such as: *How many stars are there? Which object are there 6 of?*

Teaching

- Display slide 1. Ask: *How many aliens can you see?*
- Say: *The aliens want to travel in one of these spaceships. They can only travel in the spaceship with the number that matches how many aliens there are. Can you spot the spaceship that this group of aliens can fly away in?*

- Choose a child to come and point to the correct spaceship. Ask: *Is Sofia right? Let's see.*
- Display slide 2 to show the correct spaceship. *Yes, there are 4 aliens and this spaceship shows the number 4.*
- Repeat above for the remaining slides.

Explore in groups

Pairs

- Provide sets of cards 1–10 from PCMs 1 (objects) and 5 (numerals). Encourage children to play Pairs (see Generic games rules on page 220). Each pair or group will need one set of objects cards and one set of numeral cards.

Variation: use cards from PCMs 2, 3 or 4 instead of PCM 1.

Counting collections station

- Set up collections of resources, one for each number 1 to 10. For example: 5 small-world people in a dolls house; 7 small-world farm animals in a field (green sheet of paper); 9 small-world vehicles in a car park (grey sheet of paper); 3 small-world sea creatures in bowl of water; 6 beads on a lace; 2 shoes in a box; 10 counters in a pot; a tower of 8 interlocking cubes; 4 pegs on a pegboard; 1 piece of plastic fruit on a plate. Place a set of numeral cards 1–10 from PCM 5 on the table. Encourage children to count each collection and place the matching numeral card beside the resources.

Sorting cards

- Label 10 tissue boxes (or similar) with numbers 1 to 10. Place them next to a selection of cards 1–10 from PCMs 1–4. Invite children to sort the cards by 'posting' each card into the correct box.

Session 6: Use marks and numerals to represent groups of objects

Children continue to use marks and the numerals 1 to 10 to represent the number of objects in a group.

You will need:

Maths Foundation Reading Anthology; selection of different small-world resources and counting apparatus; whiteboard; small pieces of paper; pencils; small pots; PCM 3; *Maths Foundation Activity Book C*

Getting started

- Show pages 12 and 13 (The fruit stall) of the *Reading Anthology*. Discuss the picture with the children. Point to different fruit. Ask children to tell you how many there are.

Teaching

- Show a set of 7 small-world resources or counting apparatus, e.g. cubes. Ask: *Can you help me to count these?* As you point to each cube, draw one large tally mark on the board as children say the number in the count. Then ask: *How many cubes are there?*
- Ask: *How else could I show how many cubes there are?* Lead the discussion towards using numerals. Write '7' beside the 7 tally marks.
- Repeat for another set of resources up to 10. This time, draw small dots or rings. Discuss how you have used a different way to record the number of objects in the group. Explain that it doesn't matter how you show the amount, as long as it is simple, quick and easy to write.
- Once again write the corresponding numeral beside the marks.
- Repeat for another set of resources up to 10. If appropriate, ask a child to come to the board and record marks for each object as it is counted. You could ask a different child to write the corresponding numeral.

Explore in groups

Counting collections station

- Provide 6 or 7 pots on the table, each containing up to 10 small-world resources or counting apparatus. Place some small pieces of paper next to the pots. Invite children to count the contents of each pot, making a mark on a piece of paper for each object they count. Then suggest that they turn over the paper and write the matching numeral. Prompt children to use a different piece of paper for each pot.

Counting and representing pictorial representations

- Give each child two or three different 1–10 cards from PCM 3 (cubes) and two or three small pieces of paper. Ask children to count the cubes on each card, making a mark on a piece of paper for each cube they count. Then suggest that they turn over the paper and write the matching numeral. Prompt children to use a different piece of paper for each card.

Maths Foundation Reading Anthology (adult-led)

- Together, look at pages 12 and 13 (The fruit stall) of the *Reading Anthology*. Give each child a small piece of paper. Tell each child (or point to) a different fruit in the picture. Ask them to draw a small picture of their fruit. Ask children to count how many of their fruit they can see. For each one they count, they make a mark on their paper. When they have finished counting, ask them to write the matching numeral. If time allows, repeat. Ask children to compare their pieces of paper. Do all the pieces of paper with the same fruit have the same number of marks and the same numerals?

Maths Foundation Activity Book C (adult-led)

Page 4 – How many?

Session 7: Compare two quantities and numbers to 10

Children continue to use the words 'more' or 'fewer' to compare two quantities, and the words 'more' or 'less' to compare two numbers.

You will need:

Unit 11 Session 7 slides; PCMs 4 and 5; 1 to 6 dot dice; paper; coloured pencils or crayons; *Maths Foundation Activity Book C*

Getting started

- Display slide 1. Point to a number on the number track, e.g. 5. Ask: *Can you tell me a number that is more than 5? Can you tell me any other numbers that are more than 5?*
- Repeat. Point to another number. Ask children to say numbers that are less.

Teaching notes

- Repeat several times, asking the children to say numbers that are more or less than the given number.

Teaching

- Display slide 2. Ask: *How many gloves are there? How many flip-flops can you see? Which is more: 8 or 6?*
- Point to the numbers 8 and 6 on the number track: *8 is more than 6.* Emphasise the word 'more'.
- Repeat for slides 3 to 6. Each time, use the number track to show the position of each number in the counting sequence: *… is more than …*
- Display slide 2 again. Ask: *How many gloves? How many flip-flops? There are fewer flip-flops than gloves.* Emphasise the word 'fewer'.
- Point to numbers 8 and 6 on the number track: *6 is less than 8.* Emphasise the word 'less'.
- Repeat for slides 3 to 6. For each slide, ask children to say which piece of clothing there are 'fewer' of. Use the number track to show the position of each number in the counting sequence: *… is less than …*
- Display slide 7. Ask questions such as: *Which is less: 6 or 10? Which is more: 8 or 2?* (If appropriate, display slide 8 rather than slide 7. This is more challenging as children do not have the number track to help them compare numbers.)

Explore in groups

Comparing quantities

- Shuffle a set of cards 1–10 from PCM 4 (ten-frames). Spread the cards face up on the table. Provide a 1 to 6 dot dice. Encourage children to roll the dice, count the dots, and find all the ten-frame cards with *more* dots. Children could repeat the activity several times.

Comparing quantities and numbers

- Shuffle a set of numeral cards 1–10 from PCM 5. Spread the cards face up on the table. Provide a 1 to 6 dot dice. Encourage children to roll the dice, count the dots, and find all the number cards that are *more* than the number of dots.

Painting and drawing station

- Display slide 3. (Alternatively, print the slide and place it on the table.) Suggest that children draw *more* shorts and *fewer* t-shirts than are shown on the slide.

Maths Foundation Activity Book C (adult-led)

Page 5 – More

Session 8: Share objects into two equal groups

Children begin to share an even number of objects into two equal groups.

You will need:

Unit 11 Session 8 slides; Tree Digital Tool; selection of different small-world resources; 5 trays; 10 small pots or cups; readily available classroom resources, such as balls, buckets, beanbags, hoops, bead and laces, home corner equipment, building blocks, boxes; PCM 14; small counters; coloured pencils; *Maths Foundation Activity Book C*

Getting started

- Display slide 1. Ask: *What can you tell me about the apples on these two trees?* Lead children to the fact that each tree has the same number of apples. Emphasise the phrase 'the same'.
- Repeat for slide 2. Emphasise that these two trees do not have the same number of apples. Ask: *Which tree has more apples? Which has fewer?* Emphasise the words 'more' and 'fewer'.
- Repeat for slides 3 and 4.

Teaching

- Display the Tree Digital Tool. Set it to show 2 trees. Place 6 apples below the 2 trees.
- Count the 6 apples with the children.
- Say: *The 6 apples fell from these 2 trees. The same number of apples fell from each tree.*
- Then slowly move the apples onto the 2 trees: place 1 apple on the first tree, 1 apple on the second tree, another apple on the first tree, and so on. Each tree should have 3 apples on it.

- Ask: *How many apples are there on each tree? I have shared these 6 apples equally between these 2 trees. The apples are shared fairly between the 2 trees.* Emphasise the words 'shared', 'equally' and 'fairly'.
- Click 'Clear all'. Place 4 apples below the 2 trees. Repeat as above.
- Repeat for 8 apples.
- Display slides 1 to 4 again. For each slide, ask the children to put their thumbs up if the apples have been shared equally between the 2 trees, or thumbs down if not.

Explore in groups

Counting collections station

- Set out 5 trays. On each tray, place 2 pots or cups. Place a different number of small-world resources on each tray: 2, 4, 6, 8 and 10. Encourage children to share each set of small-world resources equally between the two pots.

Sharing

- Set tasks according to the resources available (either in the classroom or outdoors). For each one, children should be able to share objects into 2 equal groups. For example, ask children to: share 10 balls between 2 buckets; share 8 beanbags between 2 hoops; share 6 beads between 2 laces; share 4 plastic fruit between 2 plates; share 2 building blocks between 2 boxes.

Ladybird spots

- Before the activity, prepare copies of PCM 14. On one copy of the sheet, draw spots (4, 6, 8 and 10) below each ladybird. Then photocopy this sheet. Cut out the ladybird cards. You will need at least 3 cards for each child. Place the cards and a pile of counters on the table. Invite children to count the spots below the ladybird

and take the same number of counters. Prompt them to share the counters equally between the two wings. They could then draw the spots on the wings.

Maths Foundation Activity Book C (adult-led)

Page 6 – Equal sharing

Assessment opportunities

Assess children's learning against the objectives for this unit, using the guidance on formative assessment on pages 24–25, and record your observations in the Unit 11 progress tracking grid on page 31. The relevant pages of *Activity Book C* can also be used for assessment.

Can the children:

- demonstrate increased understanding of the five key counting principles and of conservation of number?
- confidently recite the number names 1 to 10 in order, counting forwards and backwards, starting from any number from 1 to 10?
- say which number comes next and which number comes before when counting forwards?
- count, with increased accuracy, up to 10 objects and things that cannot be touched?
- identify and write the correct numeral to represent a set of 1 to 10 objects?
- visually recognise a quantity of 6 or fewer: *subitising*?
- count out a set of objects of a specified size from 1 to 10 from a larger group?
- recognise and write (as numerals) numbers 1 to 10 with increasing accuracy?
- use the words 'more' or 'fewer' to compare two quantities from 1 to 10?
- use the words 'more' or 'less' to compare two numbers from 1 to 10?
- share 2, 4, 6, 8 or 10 objects into two equal groups?

Unit 12 Subtraction as taking away

Theme: What is it made from?

Overview

There are several different ways of looking at subtraction. The two most commonly taught first are:

1) when a quantity is taken away from a given quantity and subtraction is used to work out how many are left – taking away

2) when a given quantity is decreased by another quantity –counting back.

This unit focuses on the first structure of subtraction: taking away. The key language in this unit includes: 'take away', 'How many are left?', 'How many are not?' and 'How many do not?' This unit also introduces children to the terms 'subtract' and 'is equal to'.

Learning objectives

Number – Understanding addition and subtraction	• **4b In practical activities and discussions begin to use the vocabulary involved in subtraction: take away** [and counting back].

Learning objectives (or parts of objectives) in bold are taught for the first time in this unit.

[This part of the learning objective is not taught in this unit.]

Vocabulary

number, count, one, two, three, …, eight, nine, ten, take away, how many are left? how many are not? how many do not? subtract, leaves, is equal to

Making connections

English: What is it made of?

Science: Materials

Preparation

You will need:

- *Maths Foundation Activity Book C,* pages 7–10.
- Unit 12 slides.
- PCMs 1, 5, 14 and 16.
- Number fluency games and activities: *Show me* (2) (page 198).
- Rhymes and songs: *Alice the camel* (7), *Five currant buns* (11), *Five little speckled frogs* (12), *Ten green bottles* (18) (pages 210–215).
- Large 1–10 number bunting and/or large number track.
- Paper, pencils, coloured pencils.

- Glue.
- 7 chairs.
- Interlocking cubes.
- Small counters.
- Readily available classroom resources, e.g. small-world resources, beads and laces, sorting trays (or egg cartons), pegboard and pegs, washing line, pegs and socks (or similar).
- Trays, small pots or cups.
- '2-3' counters (counters with '2' written on one side, and '3' on the other).

- '4-5' counters (counters with '4' written on one side, and '5' on the other).
- 8 hats (or similar).
- Large hoop.
- 10 beanbags.
- Bucket.
- Large and small 1 to 6 dot dice.
- Skittles and ball.
- Playdough station: playdough.

Before starting

Display number bunting or a large number track showing numerals 1 to 10. Each numeral should have a matching picture (e.g. 7 dots below the number 7). Refer to it regularly throughout the unit.

Children will get the most out of this unit if they already have experience of:

- counting groups of up to 10 objects
- visually recognising a quantity of 6 or fewer: *subitising*
- recording the number of objects in a group by making marks, drawing pictures or by writing numbers.

Common difficulties

Children may have difficulty remembering the original quantity before something was removed from the set. Use concrete materials, but choose objects that will leave some 'evidence' to remind the children of the original quantity. For example, use sweets in wrappers, leaving empty wrappers where sweets have been 'taken away'.

Session 1: Take away 1 object

Children are introduced to the concept of subtraction as taking away. They remove 1 object from a set and count to find out how many are left. Children are also introduced to the words: 'take away' and 'left'.

You will need:

7 chairs; interlocking cubes; PCMs 5 and 16; readily available classroom resources, e.g. small-world resources, beads and laces, sorting trays (or egg cartons), pegboard and pegs, washing line, pegs and socks (or similar); small counters

Getting started

- Place 5 chairs in a row in front of the children.
- Sing the action song *Ten green bottles* (18), to practise counting backwards from 10 to 1.

Teaching

- Ask 5 children to come and sit on the chairs. Ask: *How many children are sitting on a chair?* If possible, encourage children to 'see' how many there are without counting: *subitising*.
- Take the hand of 1 child and move them from their chair to stand by you. Ask: *Now how many children are sitting on the chairs?* Again, encourage children to answer without counting. Say: *There were 5 children sitting on the chairs. I took away 1 child, and now there are 4 children left.* Emphasise the words 'took away' and 'left'.
- Say: *5 take away 1 leaves 4.* Emphasise the words 'take away' and 'leaves'.
- Ask the 5 children to sit back with the other children.
- Remove a chair and ask 4 children to sit on the remaining chairs. Repeat above. Then say a corresponding subtraction statement: *4 take away 1 leaves 3.*
- Repeat above, removing one child from 3, 6 and 7 chairs.

Explore in groups

Take away 1 cube (adult-led)

- Place a tray of interlocking cubes on the table. Give each child a numeral card 2 to 10 from PCM 5. Place all the unused 1 to 10 number cards on the table. Ask each child to build a cube tower to match the number on their card. Then ask them to each remove 1 cube, and say a corresponding subtraction statement. Each child then looks for the number card that matches the number of cubes in their tower now.

Teaching notes

Take away 1 object

- Before the activity, prepare groups of 2 to 10 readily available classroom resources (e.g. 9 beads on a lace; 7 pegs on a pegboard; 4 socks pegged to a washing line; 10 counters in a pot; 2 shoes in a box; 6 'eggs' in an egg carton; 5 small-world animals on a green sheet of paper; a tower of 8 interlocking cubes; 3 small-world sea creatures in a bowl of water). Invite children to count the objects in each group, then remove 1 and say a corresponding subtraction statement.

Take away 1 counter

- Provide small counters and a pile of cards from PCM 16. Invite children to take a card and place a counter on top of each counter (circle) on the card. Then ask them to remove 1 counter and say a corresponding subtraction statement.

Session 2: Record subtraction (take away 1)

Children consolidate their understanding of subtraction as taking away. They record the result using pictures and numbers. Children are also introduced to the phrase: 'how many are left?'

You will need:

Unit 12 Session 2 slides; cards 2–10 from PCM 1; PCMs 14 and 16; glue; coloured pencils; *Maths Foundation Activity Book C*; pencils

Getting started

- Use the Number fluency activity *Show me* (2), to practise using fingers to represent quantities to 10.

Teaching

- Display slide 1. Ask children to tell you how many eggs there are. If possible, encourage them to 'see' how many there are without counting.
- Display slide 2 to remove one of the eggs from the carton. Say: *Jada took 1 of the eggs so that she could have a boiled egg for lunch. Now how many eggs are left?* Emphasise the phrase 'how many are left?'.

- Again, encourage children to 'see' the answer without counting. Say: *Hold up your fingers to show me how many eggs are left.*
- Display slide 3 showing the '5' number card. Say: *There were 6 eggs in the carton, 1 has gone and now there are only 5. 6 take away 1 leaves 5.* Emphasise the words 'take away' and 'leaves'.
- Repeat for slides 4 to 18. For each image, tell a suitable story involving taking away 1.

Explore in groups

Take away 1 object

- Provide multiple copies of cards 2–10 from PCM 1. Encourage children to take two or three cards. For each card, they count the objects and record the number on the card (using a numeral or marks). They then draw a line through 1 of the objects. On the other side of the card, they record how many objects are 'left'.

Variation: use cards from PCM 16 instead.

Take away 1 spot

- Before the activity, glue pairs of ladybird cards from PCM 14 back-to-back. On one side of each card, below the ladybird, write a number from 2 to 10. Encourage children to take two or three of the double-sided cards. For each card, they look at the number and draw that many spots on the ladybird. They then turn over the card and draw 1 spot fewer. Below the ladybird they record how many spots there are.

Outdoor play (adult-led)

- For children who are not ready to start recording the results of subtraction, give them more time to act out taking away 1. Take them outdoors and encourage them to pretend to be a bus (children stand one behind the other and walk in a line). They move around the area, and at each 'stop' 1 passenger gets off. When this happens, they all call out the number of passengers left.

Maths Foundation Activity Book C (adult-led)

Page 7 – Take away 1

When children have completed the page, ask them to say a corresponding subtraction statement for each set, e.g. *4 take away 1 leaves 3.*

Session 3: Take away 2 or 3 objects

Children consolidate their understanding of subtraction as taking away. They remove 2 or 3 objects from a set and count to find out how many are left. Children are also introduced to the word 'subtract' and in the context of subtraction 'equal'.

You will need:

cards 2–10 from PCM 1; PCMs 5 and 16; 8 hats (or similar); interlocking cubes; readily available classroom resources, e.g. small-world resources, beads and laces, sorting trays (or egg cartons), pegboard and pegs, washing line, pegs and socks (or similar); small counters; small pots or cups; '2-3' counters (counters with '2' written on one side, and '3' on the other)

Getting started

- Shuffle cards 2–10 from PCM 1. Place the cards face down in a pile. Turn over the top card (e.g. 6 cars) and show it to the children. Ask them to count the objects: *How many cars are there?* Then ask: *What is 6 take away 1?*
- Repeat for the other cards, asking children to take away 1 from each group of objects.

Teaching

- Ask 5 children to come and stand in a row. Give each child a hat (or similar). Ask: *How many children are wearing a hat?* If possible, encourage children to 'see' how many there are without counting.
- Then remove the hats from 2 of the children. Ask: *Now how many children are left wearing a hat?* Again, encourage children to answer without counting. Say: *There were 5 children wearing hats, I took away 2 hats and now there are 3 children left wearing hats.* Emphasise the words 'took away' and 'left'.
- Say: *5 take away 2 leaves 3.* Emphasise the words 'take away' and 'leaves'. *We can also say 5 subtract 2 is equal to 3.* Emphasise the words 'subtract' and 'equal'.
- Ask the 5 children to sit back with the other children.
- Then ask 6 children to come and stand in a row. Give each of them a hat. Repeat above, removing a hat from 2 or 3 of the children.

Then say a corresponding subtraction statement, e.g. *6 take away 3 leaves 3* or *6 subtract 2 is equal to 4.*
- Repeat above, with 5 to 8 children in a row, and removing the hats from 2 or 3 children.

Explore in groups

Take away 2 or 3 cubes (adult-led)

- Place a tray of interlocking cubes on the table. Give each child a numeral card 4–10 from PCM 5. Place all the unused 1 to 10 number cards on the table. Ask each child to build a cube tower to match the number on their card. They ask them to each remove 2 cubes, and say a corresponding subtraction statement. Each child then looks for the number card that matches the number of cubes in their tower now.

Variation: ask children to remove 3 cubes from their tower.

Take away 2 or 3 objects

- Before the activity, prepare groups of 4 to 10 readily available classroom resources (e.g. 8 beads on a lace; 5 pegs on a pegboard; 4 socks pegged to a washing line; 9 counters in a pot; 6 'eggs' in an egg carton; 7 small-world animals on a green sheet of paper; a tower of 10 interlocking cubes). Also provide several small pots or cups, each containing a '2-3' counter. Invite children to take a pot, and choose a group of resources. They count the objects, then tip the '2-3' counter out of the pot, and remove the corresponding number of objects. They then count how many objects are left, and say a corresponding subtraction statement.

Take away 2 or 3 counters

- Provide a pile of cards 5–10 from PCM 16, small counters, and several small pots or cups each containing a '2-3' counter. Invite children to take a card and place a counter on top of each counter (circle) on the card. They then tip the '2-3' counter out of a pot, and remove the corresponding number of counters from the card. They then count how many counters are left, and say a corresponding subtraction statement.

Teaching notes

Session 4: Record subtraction (take away 2 or 3)

Children consolidate their understanding of subtraction as taking away. They record the result using pictures and numbers.

You will need:

Unit 12 Session 4 slides; cards 4–10 from PCM 1; PCMs 14 and 16; small pots or cups; '2-3' counters (counters with '2' written on one side, and '3' on the other); glue; coloured pencils; beanbags; bucket; *Maths Foundation Activity Book C*; pencils

Getting started

- Sing the action song *Five little speckled frogs* (12), to practise counting backwards from 5 to 1.

Teaching

- Display slide 1. Ask children to tell you how many people there are in the bus. If possible, encourage them to 'see' how many there are without counting.
- Display slide 2 to remove 2 of the people from the bus. Say: *Two people got off the bus. How many people are left on the bus?* Emphasise the phrase 'how many are left'.
- Again, encourage children to answer without counting. Say: *Hold up your fingers to show me how many people are now on the bus.*
- Display slide 3 showing the '3' number card. Say: *There were 5 people on the bus, 2 got off and now there are 3. 5 take away 2 is equal to 3.* Emphasise the words 'take away' and 'equal'.
- Repeat for slides 4 to 24. For each image tell a suitable story involving taking away 2 or 3.

Explore in groups

Take away 2 or 3 counters

- Provide multiple copies of the cards from PCM 16, and several pots or cups each containing a '2-3' counter. Encourage children to take two or three cards. For each card, they count the counters (circles) and record the number on the card. They then tip the '2-3' counter out of a pot and draw a line through that many counters (circles) on the card. On the other side of the card, they record how many counters (circles) are 'left'.

Variation: use cards 4–10 from PCM 1 instead.

Take away 2 or 3 spots

- Before the activity, glue pairs of ladybird cards from PCM 14 back-to-back. On one side of each card, below the ladybird, write a number from 4 to 10. Provide several small pots or cups, each containing a '2-3' counter. Encourage children to take two or three of the double-sided cards. For each card, they look at the number and draw that many spots on the ladybird. They then tip a '2-3' counter out of a pot. On the other side of the card, they draw 2 or 3 fewer spots. Below the ladybird they record how many spots there are.

Outdoor play (adult-led)

- For children who are not ready to start recording the results of subtraction, give them more time to act out taking away 2 and 3. Take children outdoors and encourage them to play in groups. One child is the 'thrower'. They decide how many beanbags to put in the bucket. They tell the other children how many there are. They then gently throw beanbags to 2 or 3 other children for them to catch. All the children call out how many beanbags are left in the bucket. They count to check.

Maths Foundation Activity Book C (adult-led)

Page 8 – Take away 2 or 3

When children have completed the page, ask them to say a corresponding subtraction number sentence for each set, e.g. *7 subtract 2 is equal to 5.*

Session 5: Take away 4 or 5 objects

Children consolidate their understanding of subtraction as taking away. They remove 4 or 5 objects from a set and count to find out how many are left.

You will need:

cards 4–10 from PCM 1; large hoop; 10 beanbags; PCMs 5 and 16; interlocking cubes; readily available classroom resources, e.g. small-world resources, beads and laces, sorting trays (or egg cartons), pegboard and pegs; small counters; small pots or cups; '4-5' counters (counters with '4' written on one side, and '5' on the other)

Getting started

- Shuffle cards 4–10 from PCM 1. Place the cards face down in a pile. Turn over the top card (e.g. 5 bananas) and show it to the children. Ask: *How many bananas are there?* Then ask: *What is 5 take away 3?*
- Repeat for the other cards, asking children to subtract 1, 2 or 3 from each group of objects.

Teaching

- Place a large hoop on the floor. One at a time, place 7 beanbags into the hoop, asking children to count them as you do so. Ask: *How many beanbags are there?*
- Then say: *I'm going to take away some of these beanbags from the hoop. I want you to count how many I take away.*
- One-by-one remove 4 beanbags from the hoop, placing each one just outside the hoop. Ask: *How many beanbags did I take out of the hoop?* Encourage children to 'see' how many there are without counting.
- Then point to the beanbags that are left in the hoop. Ask: *How many beanbags are left in the hoop?* Again, encourage children to answer without counting. Say: *To begin with there were 7 beanbags in the hoop. I took away 4 beanbags and now there are 3 beanbags left in the hoop.* Emphasise the words 'took away' and 'left'.
- Say: *7 take away 4 leaves 3.* Emphasise the words 'take away' and 'leaves'. Say: *We can also say 7 subtract 4 is equal to 3.* Emphasise the words 'subtract' and 'equal'.
- Repeat several times, placing 6 to 10 beanbags in the hoop and removing 4 or 5 beanbags each time.

Explore in groups

Take away 4 or 5 cubes (adult-led)

- Repeat the *Take away 2 or 3 cubes* activity from Session 3 (page 133). Adapt by using numeral cards 6–10 from PCM 5, and asking children to take away 4 or 5 cubes.

Take away 4 or 5 objects

- Repeat the *Take away 2 or 3 objects* activity from Session 3 (page 133). Adapt by using '4-5' counters and groups of 6 to 10 resources.

Take away 4 or 5 counters

- Repeat the *Take away 2 or 3 counters* activity from Session 3 (page 133). Adapt by using '4-5' counters and cards 6–10 from PCM 16.

Session 6: Record subtraction (take away 4 or 5)

Children consolidate their understanding of subtraction as taking away. They record the result using pictures and numbers. Children are also introduced to the phrases 'How many are not?' and 'How many do not?'

You will need:

Unit 12 Session 6 slides; cards 6–10 from PCM 1; PCMs 14 and 16; small pots or cups; '4-5' counters (counters with '4' written on one side, and '5' on the other); glue; coloured pencils; playdough; *Maths Foundation Activity Book C*; pencils

Getting started

- Say the action rhyme *Five currant buns* (11), to practise counting backwards from 5 to 1.

Teaching

- Display slide 1. Ask children to tell you how many candles there are. If possible, encourage them to 'see' how many there are without counting.
- Display slide 2 to show that 4 of the candles are not lit. Ask: *How many candles are not lit?* Emphasise the phrase 'how many are not'.
- Say: *6 candles were lit then 4 of the candles were blown out. How many candles are still lit?*
- Again, encourage children to answer without counting. Say: *Hold up your fingers to show me how many candles are still lit.*
- Display slide 3 showing the '2' number card. Say: *There were 6 lit candles. 4 of the candles were blown out. So now there are only 2 candles left that are lit.*
- Say: *6 take away 4 is equal to 2.* Emphasise the words 'take away' and 'equal'. *We can also say 6 subtract 4 is equal to 2.* Emphasise the word 'subtract'.
- Repeat for slides 4 to 24. For each image tell a suitable story involving taking away 4 or 5 and using the phrase 'How many are not?' or 'How many do not?'.

Teaching notes

Explore in groups

Take away 4 or 5 counters

- Repeat the *Take away 2 or 3 counters* activity from Session 4 (page 134). Adapt by using '4-5' counters.

Variation: use cards 6–10 from PCM 1 instead.

Take away 4 or 5 spots

- Repeat the *Take away 2 or 3 spots* activity from Session 4 (page 134). Adapt by writing a number from 6 to 10 on one side of the ladybird, and using '4-5' counters.

Playdough station (adult-led)

- For children who are not ready to start recording the results of subtraction, give them more time to act out taking away 4 and 5. Let them play with playdough. Ask them to make a set of at least 6 balls. Then ask them to choose 4 or 5 of the balls and squash them flat. Can they say how many balls they started with, how many they squashed, and how many are left unsquashed?

Maths Foundation Activity Book C (adult-led)

Page 9 – Take away 4 or 5

When children have completed the page, ask them to say a corresponding subtraction number sentence for each set, e.g. *7 subtract 4 leaves 3.*

Session 7: Take away up to 6 objects

Children consolidate their understanding of subtraction as taking away. They remove up to 6 objects from a set and count to find out how many are left.

You will need:

interlocking cubes; large and small 1 to 6 dot dice; PCM 5 and 16; readily available classroom resources, e.g. small-world resources, beads and laces, sorting trays (or egg cartons), pegboard and pegs; small counters; small pots or cups

Getting started

- Give each child a tower of 7, 8, 9 or 10 interlocking cubes. Ask children to count the cubes in their tower. Say: *Stand up if you have a tower of 8 cubes.* Check, and ask all these children to sit together. Repeat for towers of 7, 9 and 10 cubes so that children are sitting in four groups.
- Pointing to each group in turn say: *So, everyone in this group has a tower of 7 cubes. Everyone over here has 8 cubes.*

Teaching

- Roll the large 1 to 6 dot dice and say the number, e.g. 5. Tell children that you want them to take away that many cubes from their tower and place them on the floor in front of them.
- Say to the 9-cube group: *You all started with 9 cubes in your tower. You have just taken 5 cubes from your tower. How many cubes are left in your tower?* If possible, encourage the children to 'see' how many there are without counting. Say: *9 take away 5 leaves 4.*
- Repeat for the other three groups.
- Then tell all the children to remake their towers.
- Confirm again how many cubes are in each groups' towers then roll the dice and repeat above. Each time, confirm the answer and say a corresponding subtraction statement.
- Repeat several times.

Explore in groups

Take away up to 6 cubes (adult-led)

- Repeat the *Take away 2 or 3 cubes* activity from Session 3 (page 133). Adapt by using numeral cards 7–10 from PCM 5, and asking children to take away any number of cubes from 1 to 6.

Take away up to 6 objects

- Repeat the *Take away 2 or 3 objects* activity from Session 3 (page 133). Adapt by using small 1 to 6 dot dice and groups of 7 to 10 resources.

Take away up to 6 counters

- Repeat the *Take away 2 or 3 counters* activity from Session 3 (page 133). Adapt by using small 1 to 6 dot dice and cards 7–10 from PCM 16.

Session 8: Record subtraction (take away up to 6)

Children consolidate their understanding of subtraction as taking away. They record the result using pictures and numbers.

You will need:

Unit 12 Session 8 slides; cards 7–10 from PCM 1; PCMs 14 and 16; small pots or cups; small 1 to 6 dot dice; glue; coloured pencils; skittles and ball; *Maths Foundation Activity Book C*; pencils

Getting started

- Sing the song *Alice the camel* (7), to practise counting backwards from 5 to 1.

Teaching

- Display slide 1. Ask children to tell you how many children there are in the dodgem cars. Say: *Let's count them. 1, 2, 3, … 9.*
- Display slide 2 to remove 3 children from their cars. Say: *3 children got off the dodgems. How many children are left?* Emphasise the phrase 'how many are left?'. Say: *Hold up your fingers to show me how many children are left on the dodgems?*
- Display slide 3 showing the '6' number card. Say: *There were 9 children in the dodgem cars, 3 got off and now there are 6. 9 subtract 3 is equal to 6.* Emphasise the words 'subtract' and 'equal'.
- Repeat for slides 4 to 24. For each image tell a suitable story involving subtracting up to 6. Use the appropriate language such as 'How many are left?', 'How many are not?' or 'How many do not?'.

Explore in groups

Take away up to 6 counters

- Repeat the *Take away 2 or 3 counters* activity from Session 4 (page 134). Adapt by using cards 7–10 from PCM 16, and using small 1 to 6 dot dice.

Variation: use cards 7–10 from PCM 1 instead.

Take away up to 6 spots

- Repeat the *Take away 2 or 3 spots* activity from Session 4 (page 134). Adapt by writing a number from 7 to 10 on one side of the ladybird, and using small 1 to 6 dot dice.

Skittles (adult-led)

- For children who are not ready to start recording the results of subtraction, give them more time to act out taking away up to 6. Let them play outdoors with 10 skittles and a ball. Encourage them to take turns to choose how many skittles to set up, then roll the ball to knock some down. Can they say how many skittles they started with, how many they knocked down, and how many are left standing?

Maths Foundation Activity Book C (adult-led)

Page 10 – Take away

When children have completed the page, ask them to say a corresponding subtraction statement for each set.

Assessment opportunities

Assess children's learning against the objectives for this unit, using the guidance on formative assessment on pages 24–25, and record your observations in the Unit 12 progress tracking grid on page 32. The relevant pages of *Activity Book C* can also be used for assessment.

Can the children:

- count up to 10 objects?
- visually recognise a quantity of 6 or fewer: *subitising*?
- recognise and write (as numerals) numbers 1 to 10?
- demonstrate an understanding of subtraction as take away?
- use the vocabulary associated with subtraction as take away: 'take away', 'How many are left?', 'How many are not?', 'How many do not?', 'subtract' and 'is equal to'?
- begin to record a subtraction using pictures or numbers?

Unit 13 Subtraction as counting back

Theme: People at work

Overview

There are several different ways of looking at subtraction. The two most commonly taught first are:

1) when a quantity is taken away from a given quantity and subtraction is used to work out how many are left – taking away

2) when a given quantity is decreased by another quantity – counting back.

This unit focuses on the second structure of subtraction: counting back. The key language in this unit includes: 'start at and count back', 'start, then, now'. This unit also consolidates language that was introduced in Unit 12: 'subtract' and 'is equal to', and introduces the term 'minus'.

Children apply their understanding of counting back to find 1 less than a number from 2 to 10.

Learning objectives

Number – Understanding addition and subtraction	• *4b In practical activities and discussions begin to use the* *vocabulary involved in subtraction: [take away and]* **counting back.** • *4c Find [1 more and]* **1 less** *than a number from 1 to 10.*

Learning objectives in italics have been taught previously. In this unit they are consolidated and/or extended.

Learning objectives (or parts of objectives) in bold are taught for the first time in this unit.

[This part of the learning objective is not taught in this unit.]

Vocabulary

number, count, one, two, three, …, eight, nine, ten, start at, count backwards, count back, 1 less, start, then, now, subtract, minus, left, leaves, is equal to

Making connections

English: People at work

Science: Taking care of our world

Preparation

You will need:

- *Maths Foundation Activity Book C*, pages 11–14.
- Unit 13 slides.
- Digital Tool: Number track.
- PCM 2 (cards 1–5 only).
- PCM 3 (cards 2–10 only).
- PCM 4 (cards 6–10 only).

- PCMs 5, 15 and 16.
- Number fluency games and activities: *Finger counting* (1), *Show me* (2) (page 198).
- Rhymes and songs: *Alice the camel* (7), *Zoom, zoom, we're going to the moon* (8), Five little monkeys (10), *Five little*

speckled frogs (12), *Ten in the bed* (17) (pages 210–215).
- Large 1–10 number bunting and/or large number track.
- Paper, pencils, coloured pencils.
- Interlocking cubes.
- Trays, small pots or cups.
- Chalk.
- '1-2' counters (counters with '1' written on one side, and '2' on the other).
- '3-4' counters (counters with '3' written on one side, and '4' on the other).
- '4-5' counters (counters with '4' written on one side, and '5' on the other).
- 6 chairs.
- Small counters.
- Large 1 to 6 dot dice (with the 1-dot face covered with paper and a 7-dot pattern drawn on).
- Washing line and pegs.
- Toy cars.

- 7 small pieces of paper (ideally black or grey).
- Large hoop.
- 8 beanbags.
- Counting collections station: a selection of different small-world resources (e.g. people at work), plates or dishes.
- Sand station: sand tray/pit, small-world people, e.g. people at work.

Before starting

Display number bunting or a large number track showing numerals 1 to 10. Each numeral should have a matching picture (e.g. 5 dots below the number 5). Refer to it regularly throughout the unit.

Before teaching this unit, look back at the Assessment section at the end of Unit 12 (page 137), to identify the prerequisite learning for this unit.

Common difficulties

When subtracting by counting back, children must not only count back (with numbers decreasing as they count). They must also count how many they have counted back (which increases as they count). For example, as children count back 3 from 8, they are counting back – *7, 6, 5* – while remembering how many they have counted – 1, 2, 3. Children often find it difficult to keep track of how many steps back have been taken. Encourage them to develop strategies such as finger counting or using mark-making to support them.

Another common error is children counting the starting number. For example, when working out 5 subtract 3, they count: *5, 4, 3*, therefore getting the wrong answer. It is vital to model how to count backwards by 'putting the start number in our head' and **then** counting backwards.

Session 1: Count backwards, starting from any number to 10

Children count backwards, starting from, and ending at, any number from 1 to 10.

You will need:

cards 6–10 from PCMs 4 and 5; PCM 15 ; large 1–10 number bunting and/or large number track

Getting started

- Ensure that all children can clearly see the 1–10 number track/bunting.
- Use the Number fluency activity *Finger counting* (1), to practise counting forwards from 1 to 10 and backwards from 10 to 1.

Teaching

- Point to '3' on the number track. Say: *We're going to count back to 10 again, but this time we're not going to start from 10. We're going to count back from 7. Ready? Let's go!* Point to 7 on the number track and count back together along the track to 1.

Teaching notes

- Repeat several times, counting back to 1 from a number other than 10. Occasionally, ask children to count back to a number other than 1, e.g. from 8 to 3.
- Without referring to the number track, occasionally ask questions such as: *What's the next number in this count: 6, 5, 4, …? Tell me the number that comes next: 8, 7, 6, 5, …*
- Show children a numeral card 6–10 from PCM 5. Ask children to count back from that number to 1 (or another number). Again, point to the numbers on the track as children count.
- Repeat several times.

Explore in groups

Repeat *Teaching* (adult-led)

- Repeat the *Teaching* activity with smaller groups of children. Give each child a 1–10 number track from PCM 15. Encourage them to count back, starting from any number from 1 to 10, pointing to the starting number on the track and counting back along the track to 1, or another number.

Number track counting

- Shuffle a set of cards 6–10 from PCM 4 (ten-frames) and place them face down in a pile. Also provide 1–10 number tracks from PCM 15. Invite children to choose a card, count the dots and place their finger on that number on the number track. Then ask them to count back along the track to 1.

Counting back to 1

- Shuffle two sets of numeral cards 6–10 from PCM 5. Place them face down in a pile. Encourage children to play in pairs. They take turns to turn over the top card, say the number, then count back to 1. They repeat until all 10 cards have been used.

Session 2: Subtraction as counting back (A)

Children are introduced to subtraction as counting back. They consolidate their understanding of the words 'subtract' and 'is equal to'.

You will need:

towers of 5 and 6 interlocking cubes; plates or dishes; selection of different small-world resources; small pots or cups; '1-2' counters (counters with '1' written on one side, and '2' on the other)

Getting started

- Say the action rhyme *Zoom, zoom, we're going to the moon* (8), to practise counting backwards from 5 to 1.

Teaching

- Hold up a tower of 5 cubes. Ask: *How many cubes are in this tower?* If possible, encourage children to 'see' how many there are without counting: *subitising*.
- Say: *There are 5 cubes in this tower. I'm going to remove 2 cubes from the tower. 5,* [remove 1 cube] *4,* [remove another cube] *3.*
- Say: *I started with 5 cubes. Then I took off 2 cubes* [hold up the 2 cubes] *and now there are 3 cubes left* [hold up the tower of 3 cubes]. Emphasise the words 'started', 'then', 'now' and 'left'. Say: *5 subtract 2 is equal to 3.*
- Repeat, using towers of 5 and 6 interlocking cubes, and removing 1 or 2 cubes each time.

Explore in groups

Counting collections station (adult-led)

- On the table, place a selection of small-world resources of the same type, e.g. people at work. Give each child a plate/dish. Ask children to put 6 resources on their plate. Then ask them to remove 2 resources from their plate, one-by-one, counting back as they do so. For example: *6,* [remove 1 resource] *5,* [remove another resource] *4.* Then ask each child to say a corresponding statement: *I started with 6, then I took out 2 and now there are 4 left.* Repeat several times, asking children to put 5 or 6 resources on their plate and to remove 1 or 2 resources each time.

Counting back from 10

- Provide a tower of 10 interlocking cubes. Invite children to work as a group. One child takes the tower and says: *10*. They then pass it to the second child who removes a cube and says: *9*. This continues until one child is left with 1 cube.

Counting back cubes

- Provide towers of 5 interlocking cubes, and several small pots or cups each containing a '1-2' counter. Invite children to work in pairs. They each take a cube tower. They take turns to tip a '1-2' counter from a pot and remove that number of cubes from their tower. They count back as they remove the cube(s). For example, if 2 was rolled: *5*, [remove 1 cube] *4*, [remove another cube] *3*.

Variation: provide towers of 6 interlocking cubes.

Session 3: Subtraction as counting back (B)

Children consolidate their understanding of subtraction as counting back.

You will need:

6 chairs; numeral cards 1–6 from PCM 5; cards 1–3 from PCM 3; small counters; cards 5 and 6 from PCM 16; interlocking cubes

Getting started

- Place 5 chairs in a row in front of the children. Place numeral cards 1–5 from PCM 5 on the backs of the chairs in order from 1 to 5 (facing the children).
- Sing the song *Alice the camel* (7), to practise counting backwards from 5 to 1.

Teaching

- Ask 5 children to come and sit on the chairs (in front of the number cards). Ask: *How many children are sitting on a chair?* If possible, encourage the children to 'see' how many there are without counting.
- Take the hand of the child on the '5' chair and move them from their chair to stand by you. (Leave the number card on the chair.) Ask: *Now how many children are sitting on the* chairs? Again encourage children to answer without counting. Say: *We started with 5 children sitting on chairs, then I took away one child, and now there are 4 children left.* Emphasise the words 'started', 'then' and 'now'.
- Point to the number '5' on the empty chair. Say: *There were 5 children sitting on these chairs. I took away, or subtracted, one.* Count back 1 from the '5' chair to the '4' chair. *Now there are 4 children left.* Ask the child sitting on the '4' chair to stand and show the number. Say: *5 subtract 1 is equal to 4.*
- Ask the 5 children to sit back with the other children.
- Ask another 5 children to sit on the chairs. Repeat above, but this time remove 2 children from their chairs (the children sitting on the '4' and '5' chairs). Then say a corresponding subtraction number sentence: *5 subtract 2 is equal to 3.*
- Repeat above, removing 3 children from the 5 chairs.
- If time allows, repeat, removing 1, 2 or 3 children from 6 chairs.

Explore in groups

Count back 1, 2 or 3 (adult-led)

- Place a pile of counters on the table. Give each child a 5 counters card from PCM 16. Ask children to place a counter on top of each counter (circle) on their card. Then ask them to remove 2 counters, one-by-one, counting back as they do so. For example: *5*, [remove 1 counter] *4*, [remove another counter] *3*. Then ask each child to say a corresponding statement: *I started with 5 counters, then I took off 2 counters and now there are 3 counters left.* Repeat several times, asking children to remove 1 or 3 counters from their card.

Variation: give each child a 6 counters card from PCM 16 and ask them to remove 1, 2 or 3 counters.

Counting back from 10

- Provide a tower of 10 interlocking cubes. Invite children to work as a group. One child takes the tower and says: *10*. They then pass it to the second child who removes 2 cubes and says: *9, 8*. The tower is then passed to the third child who removes 2 cubes and says: *7, 6*. This continues until one child is left with 2 cubes.

Counting back cubes

- Provide towers of 5 interlocking cubes, and several shuffled sets of cards 1–3 from PCM 3 (cubes). Invite children to work in pairs. They take a set of cards and spread them out face down. They each take a cube tower. Each child turns over a card, counts the cubes on their card, and removes that number of cubes from their tower. They count back as they remove the cube(s). For example, if the 3 cube card is picked: *5*, [remove 1 cube] *4*, [remove another cube] *3*, [remove a third cube] *2*.

Variation: provide towers of 6 interlocking cubes.

Session 4: Find 1 less than a number from 2 to 10

Children consolidate their understanding of subtraction as counting back. They use the number track as a tool for finding 1 less than a given number.

You will need:

large floor 1–10 number track (if necessary, use chalk to draw a track outdoors); large 1 to 6 dot dice; sand tray; small-world resources; cards 2–10 from PCMs 3 and 5; PCM 15; interlocking cubes; *Maths Foundation Activity Book C*; pencil

Getting started

- Before the session, on the large 1 to 6 dot dice cover the 1-dot face with paper and draw a 7-dot pattern on it.
- Ensure that all children can see the large floor 1–10 number track.
- Sing the action song *Ten in the bed* (17), to practise counting backwards from 10 to 1.

Teaching

- Roll the dice, e.g. 5. Ask: *How many dots are there?* Stand on that number on the number track.

- Say: *Now I'm going to jump back 1 space along the track. What number will I land on?*
- Jump back 1 space. Say: *I have landed on 4. 4 is 1 less than 5.* Emphasise the words '1 less'.
- Repeat two or three times. Model the different ways the vocabulary can be used, e.g. *1 less than 3 is 2.*
- Ask a child to roll the dice, stand on that number on the number track and then jump back 1 space along the track. Encourage them to use the words '1 less'.
- Repeat several times with different children.
- Ask children to say the number that is 1 less than 8, 9 or 10.

Explore in groups

Sand station

- Place a selection of small-world resources (e.g. people at work) in the sand tray. Let children play, e.g. they could pretend the small-world people are builders on a building site. Encourage children to play in pairs. One child says a number from 2 to 10, e.g. 7. The other child counts out 1 fewer small-world resources. Encourage them to say, e.g. *6 is 1 less than 7.*

1 less

- Shuffle a set of numeral cards 2–10 from PCM 5. Place them face down in a pile. Encourage children to play a game in pairs. They turn over the top card and both say the number that is 1 less. The first child to say the correct answer keeps the card. The overall winner is the child with the most cards.

1 less than

- Shuffle a set of cards 2–10 from PCM 3 (cubes). Place them face down in a pile. Provide a tray of interlocking cubes and a 1–10 number track from PCM 15. Invite children to take a card. Ask them to build a tower with 1 less cube than the number on the card. Ask them to point to that number on the number track and say, e.g. *1 less than 7 is 6.*

Maths Foundation Activity Book C (adult-led)

Page 11 – 1 less

Session 5: Subtract a quantity of up to 4 from 5, 6 or 7

Children consolidate their understanding of subtraction. They use the number track as a tool for counting back. Children are also introduced to the word 'minus'.

You will need:

washing line and pegs; PCM 5; 7 toy cars; 7 small pieces of paper (ideally black or grey); small counters; PCM 15 (enlarged and normal size); toy cars (or similar); selection of different small-world resources

Getting started

- Before the session, create a washing line using the numeral cards 1–10 from PCM 5. Hang it low enough for children to remove number cards. Place 6 small pieces of black or grey paper in a row on the table.
- Use the Number fluency activity *Show me* (2), to practise using fingers to represent quantities to 10.

Teaching

- Explain to children that the pieces of black/grey paper are parking spaces in a car park.
- Count out 6 toy cars, placing each one in a 'parking space'. Ask: *How many cars are there?* Confirm the answer and unpeg the '6' card from the washing line and place it beside the cars.
- Then 'drive' 2 cars out of the car park, placing them to one side. Say: *2 cars have driven away. How many cars are left?*
- Point to the '6' number card and say: *There were 6 cars in the car park. 2 drove away.* Pointing to the washing line, count back 2 from the space where the '6' number card was: *5, 4.*
- Say: *Let's check using our fingers. There were 6 cars* [point to the '6' number card and hold up 6 fingers]. *2 cars drove away* [fold down 2 fingers]. *How many cars are left? 4. 6 subtract 2 is 4.*
- Repeat, using 5, 6 or 7 cars and removing 1, 2, 3 or 4 cars.

Explore in groups

Count back 1, 2 or 3 (adult-led)

- Place a pile of small counters on the table and give each child a 1 to 10 number track from PCM 15. Ask children to place a counter on top of each number from 1 to 5 on their number track. Then ask them to remove 3 counters, one-by-one, counting back as they do so. For example: *5*, [remove 1 counter] *4*, [remove another counter] *3*, [remove a third counter] *2*. Then ask each child to tell you how many counters are left. Tell them to lift up the last counter to confirm the answer of 2. Repeat several times, asking children to remove 1, 2 or 4 counters.

Variations: ask children to place counters on top of numbers 1 to 6, or 1 to 7, on their number tracks.

Driving away

- Provide an enlarged 1–10 number track from PCM 15, and 5 toy cars (or similar). Encourage children to work in pairs. They place the 5 cars over the numbers 1 to 5 on the number track. One child then tells the other to drive away 1, 2, 3 or 4 cars, e.g. *Drive away 3 cars.* The other child then drives that number of cars, one-by-one, from the number track, counting back as they do so. The child then says how many cars are left and lifts up the last car to confirm the answer.

Variations: ask children to use 6 or 7 cars.

Small-world

- Repeat the *Driving away* activity above, but instead of using toy cars use small-world resources such as people at work.

Session 6: Subtract a quantity of up to 4 from 5, 6, 7 or 8

Children consolidate their understanding of subtraction. They continue to use the number track as a tool for counting back. Children are also introduced to the word 'minus'.

You will need:

Number track Digital Tool; large hoop; 8 beanbags; interlocking cubes; PCM 15; small pots or cups; '3-4' counters (counters with '3'

Teaching notes

written on one side, and '4' on the other); cards 1–4 from PCM 2; small counters; *Maths Foundation Activity Book C*; pencils; coloured pencils

Getting started

- Display the Number track Digital Tool. Set it to show a 1–10 number track.
- Say the action rhyme *Five little monkeys* (10), to practise counting backwards from 5 to 1.

Teaching

- Place a large hoop on the floor. One at a time, place 6 beanbags into the hoop. Ask children to count each one as you place it inside the hoop. Then ask: *How many beanbags are there?*
- On the Number track Digital Tool, move the kangaroo to 6. Draw a circle around number 6 on the track. Say: *We start with 6 beanbags.*
- Then say: *I'm going to take some of these beanbags out of the hoop. I want you to count how many I take out.*
- One-by-one, slowly remove 4 beanbags from the hoop, placing each one just outside the hoop. Then ask: *How many beanbags did I take out of the hoop?* If possible, encourage children to 'see' how many there are without counting.
- Point to the beanbags that are left in the hoop. Ask: *Now how many beanbags are in the hoop?* Again, encourage children to answer without counting.
- Point to the kangaroo on the number track. Say: *We started with 6 beanbags, then I took out 4: 1* [draw a jump from 6 to 5], *2* [draw a jump from 5 to 4], *3* [draw a jump from 4 to 3], *4* [draw a jump from 3 to 2]. *Now there are just 2 beanbags left in the hoop.* Unclick 'Draw' and then click on 2 to highlight the answer on the track. Emphasise the words 'started', 'then' and 'now'.

- Say: *So, 6* [point to 6 on the number track] *subtract 4* [point to the 4 spaces from 6 and 2] *leaves 2* [point to 2]. *We can also say 6 minus 4 is equal to 2.* Emphasise the word 'minus'.

- Reset the Number track Digital Tool and remove the beanbags from the hoop. Repeat several times, placing 5 to 8 beanbags in the hoop and removing 1, 2, 3 or 4 beanbags.

Explore in groups

Take off cubes (subtracting 3 or 4 from 5, 6 or 7)

- Provide towers of 5 interlocking cubes, 1–10 number tracks from PCM 15 with a circle around number '5', and several small pots or cups each containing a '3-4' counter. Invite children to take a cube tower and tip a 3-4 counter out of a pot, e.g. 3. Ask them to remove that number of cubes from their tower. Encourage them to put their finger on '5' on a number track and count back 3 (or 4) and say the result. They then look at their tower to check.

Variations: provide towers of 6 (or 7) interlocking cubes, and number tracks with number 6 (or 7) circled.

Counting back fingers (subtracting up to 4 from 5, 6 or 7)

- Shuffle a set of cards 1–4 from PCM 2 (fingers). Place them face down in a pile. Provide small counters and 1–10 number tracks from PCM 15 with a circle around number '6'. Encourage children to place a counter on 6 on a number track, and take a finger card from the pile. Ask them to count back along the number track the number of fingers shown on the card. Prompt them to say a corresponding subtraction statement, e.g. *6 minus 3 makes 3.*

Variations: provide number tracks with number 5 (or 7) circled.

Maths Foundation Activity Book C (adult-led)

Page 12 – Count back 2

When children have completed the page, ask children to say a corresponding subtraction number sentence for each number track, e.g. *4 minus 2 is equal to 2.*

Session 7: Subtract a quantity of up to 4 from any number to 10

Children consolidate their understanding of subtraction. They continue to use the number track as a tool for counting back.

You will need:

Unit 13 Session 7 slides; interlocking cubes; PCM 15; small pots or cups; '3-4' counters (counters with '3' written on one side, and '4' on the other); cards 1–4 from PCM 2; small counters; *Maths Foundation Activity Book C*; pencils; coloured pencils

Getting started

- Sing the action song *Five little speckled frogs* (12), to practise counting backwards from 5 to 1.

Teaching

- Display slide 1. Ask children to tell you how many loaves of bread there are. (5). If possible, encourage them to 'see' how many there are without counting.
- Display slide 2 to remove 2 loaves. Say: *2 loaves of bread have been taken out of the baking tray. How many loaves of bread are left in the tray?*
- Again, encourage children to answer without counting. Say: *Hold up your fingers to show me how many loaves of bread are left in the tray.*
- Point to the number track. Say: *We started with 5 loaves* [point to 5 on the number track], *then 2 loaves were taken out of the tray: 1* [draw a jump from 5 to 4], *2* [draw a jump from 4 to 3]. *Now there are 3 loaves of bread left in the tray* [point to 3]. Emphasise the words 'started', 'then' and 'now'.
- Say: *5 subtract 2 is equal to 3. How else can we say this?* Praise children who suggest other sentences such as *5 minus 2 is equal to 3. 5 take away 2 leaves 3.*
- Repeat for slides 3 to 12.

Explore in groups

Take off cubes (subtracting 3 or 4 from 8, 9 or 10)

- Repeat the *Take off cubes (subtracting 3 or 4 from 5, 6 or 7)* activity from Session 6 (page 144). Adapt by providing towers of 8 (or 9 or 10) interlocking cubes, and number tracks from PCM 15 with number 8 (or 9 or 10) circled.

Counting back fingers (subtracting up to 4 from 8, 9 or 10)

- Repeat the *Counting back fingers (subtracting up to 4 from 5, 6 or 7)* activity from Session 6 (page 144). Adapt by providing number tracks from PCM 15 with number 9 (or 8 or 10) circled.

Maths Foundation Activity Book C (adult-led)

Page 13 – Count back 3 or 4

When children have completed the page, ask them to say a corresponding subtraction number sentence for each track, e.g. *7 subtract 3 is equal to 4.*

Session 8: Subtract a quantity of up to 5 from any number to 10

Children consolidate their understanding of subtraction. They continue to use the number track as a tool for counting back.

You will need:

Unit 13 Session 8 slides; interlocking cubes; PCM 15; small pots or cups; '4-5' counters (counters with '4' written on one side, and '5' on the other); cards 1–5 from PCM 2; small counters; *Maths Foundation Activity Book C*; pencils; coloured pencils

Getting started

- Use the Number fluency activity *Show me* (2), to practise using fingers to represent quantities to 10.

Teaching notes

Teaching

- Display slide 1. Ask children to tell you how many cooking utensils there are (5). If possible, encourage them to 'see' how many there are without counting. Circle number '5' on the number track.
- Point to the hand above the number track and ask children to tell you how many fingers are showing (3).
- Draw a line through 3 of the cooking utensils, saying: *1, 2, 3. The cook has taken 3 things away to cook with.*
- Then say: *There were 5 things used for cooking, and the cook took 3 of them away to use. Hold up your fingers to show me how many things are left.*
- Say: *There were 5 things to start with* [point to 5 on the number track], *then 3 of them were taken away to be used: 1* [draw a jump from 5 to 4], *2* [draw a jump from 4 to 3], *3* [draw a jump from 3 to 2]. *Now there are 2 things left* [point to 2]. Emphasise the words 'start', 'then' and 'now'.
- Say: *5 subtract 3 is equal to 2. Remember, we can also say 5 minus 3 is equal to 2, or 5 take away 3 leaves 2.*
- Repeat for slides 2 to 6.

Explore in groups

Take off cubes (subtracting 4 or 5 from 6, 7, 8, 9 or 10)

- Repeat the *Take off cubes (subtracting 3 or 4 from 5, 6 or 7)* activity from Session 6 (page 144). Adapt by providing towers of 7 (or 6, 8, 9 or 10) interlocking cubes, '4-5' counters, and number tracks from PCM 15 with the number 7 (or 6, 8, 9 or 10) circled.

Counting back fingers (subtracting up to 5 from 6, 7, 8, 9 or 10)

- Repeat the *Counting back fingers (subtracting up to 4 from 5, 6 or 7)* activity from Session 6 (page 144). Adapt by using cards 1–5 from PCM 2 (fingers) and number tracks from PCM 15 with the number 6 (or 7, 8, 9 or 10) circled.

Maths Foundation Activity Book C (adult-led)

Page 14 – Count back

When children have completed the page, ask them to say a corresponding subtraction number sentence for each track, e.g. *9 subtract 4 is equal to 5.*

Assessment opportunities

Assess children's learning against the objectives for this unit, using the guidance on formative assessment on pages 24–25, and record your observations in the Unit 13 progress tracking grid on page 32. The relevant pages of *Activity Book C* can also be used for assessment.

Can the children:

- count backwards, starting from, and ending at, any number from 1 to 10?
- count up to 10 objects?
- visually recognise a quantity of 6 or fewer: *subitising*?
- recognise and write (as numerals) numbers 1 to 10?
- demonstrate an understanding of subtraction as counting back, including using a number track?
- find 1 less than a number from 2 to 10?
- use the vocabulary associated with subtraction as counting back: 'start at', 'count back', 'subtract', 'minus', 'is equal to' and 'left'?
- begin to record a subtraction using pictures, marks or numbers?

Unit 14 Mass and capacity

Theme: Opposites/Directions

Overview

This unit introduces children to mass and capacity.

The most important aspects of measures at this stage are for children to make direct and indirect comparisons of two objects/containers, and to hear accurate vocabulary modelled.

Children will start describing and making comparison using the language of size: big, bigger, small and smaller. They are swiftly moved on to more precise language. Terms such as heavy, heavier, light and lighter are used to describe and compare masses. Terms such as full, nearly full, empty, nearly empty, 'has more' and 'has less' are used to describe and compare volumes. When comparing two capacities the terms 'holds more' and 'holds less' are used.

Children also begin to develop an awareness that the mass or capacity of an object is unchanged, even if an object is rearranged or its shape is changed: *conservation*.

Learning objectives

Geometry and Measure – Measurement	• **8b Use everyday language to describe and compare mass including heavy, heavier, light and lighter.** • **8c Use everyday language to describe and compare capacity and volume including more, less, full and empty.**

Learning objectives in bold are taught for the first time in this unit.

Vocabulary

measure, size, compare, big, bigger, small, smaller, mass, heavy, heavier, light, lighter, balance scale, balances, has the same mass (weighs the same), capacity, holds more, holds less, holds the same, volume, full, nearly full, empty, nearly empty, has more, has less, has the same

Making connections

English: Directions

Science: Pushes and pulls

Preparation

You will need:

- *Maths Foundation Reading Anthology*, pages 24 and 25.
- *Maths Foundation Activity Book C*, pages 15–19.
- PCM 27.
- Selection of objects suitable for describing and comparing sizes of two objects, but that focus on mass (including some objects that are the same mass), e.g. feathers, balloons, scraps of paper, paper clips, pencils, crayons, interlocking cubes, beads, balls, books, shoes, building blocks, bean bags, small-world resources such as people, animals and

Teaching notes

transport, soft toys, Home corner props, recycled boxes and packaging, natural materials (leaves, stones, pine cones).
- Selection of containers suitable for describing and comparing capacities and volumes (including some transparent containers, identical containers and containers with the same capacity but that are of different shapes), e.g. buckets, baskets, jugs, pots and pans, mugs, cups, glasses, measuring cups, ladles, spoons, measuring spoons, food storage containers, recycled packaging such as tins, cartons, plastic bottles, containers.
- Selection of resources suitable for filling containers, e.g. large jug of water, sand, dried lentils, pebbles, shells, marbles, centimetre cubes.
- Small pots or cups.
- Sorting hoops.
- Sorting trays or bags.
- Outdoor equipment and resources.
- Paper, pencils, coloured pencils and crayons.
- Magazines, newspapers, catalogues, etc.
- Two identical buckets or bags.
- Balance scales.

- Water tray, sand tray, large tray.
- Playdough station: playdough, balance scales.

Before starting

As part of Session 1, start to create a wall and table display of images and objects linked to mass and capacity. Throughout the unit ask children to look for examples to include in the display (either real-world examples or pictures from magazines, comics, catalogues, toy packaging etc.).

Children will get the most out of this unit if they already have some experience of:

- using everyday language to describe size, e.g. *That shirt is too big for me. That's a small cup.*
- comparing lengths, heights and widths of objects using words such as longer, shorter, taller, wider, narrower and the same.

As part of each session, try to take children outdoors so that they can experience a range of different objects associated with mass and capacity.

Common difficulties

Mass can be a tricky concept for some children, because mass is invisible, and children tend to make judgements based on what they can see. A common misconception is that the larger the size of an object, the greater the mass. For example, children may think that a beach ball is heavier than a baseball because it is larger. Provide lots of opportunities to compare two objects where the smaller one is heavier. Ask children to predict which will be heavier and to hold the objects so that they can experience for themselves the differences in mass.

Children may confuse the capacity of an object with its height. They may assume that a taller container will have a greater capacity. Demonstrate that a taller container does not necessarily hold more than a shorter one. Give children practical experience of this, allowing them to compare capacities of differently proportioned containers.

Maths background

It is important to note the difference between mass and weight, and capacity and volume. Mass refers to the amount of matter something is made up of, whereas weight is the amount of gravity acting on something. Capacity refers to the amount of space inside a container – the maximum amount it can hold. Volume however, refers to how much a container is actually holding, or the amount of space taken up by an object.

In everyday life, the terms mass and weight, and capacity and volume, are often used interchangeably. It is not expected at this stage that children use these words accurately. However, it is important that teachers model the most appropriate terminology.

Session 1: Describe and compare sizes (revision)

Children describe and compare the sizes of objects. They use the words 'big', 'bigger', 'small' and 'smaller'.

You will need:

Maths Foundation Reading Anthology; selection of objects for describing size ('big' and 'small'), but that focus on mass, e.g. feathers, balloons, scraps of paper, paper clips, pencils, crayons, interlocking cubes, beads, balls, books, shoes, building blocks, bean bags, small-world resources, soft toys, Home corner props, recycled boxes and packaging, natural materials; sorting hoops; bags

Getting started

- This session revises previous learning about size (big and small), in preparation for exploring mass in Session 2.
- Before the session, place on a table a selection of objects suitable for describing and comparing sizes.
- Remind children of the words 'big' and 'small'. Hold up an object such as a soft toy. Say, for example: *I'm big. Bear is small.* Emphasise the words 'big' and 'small'.
- Point to, or hold up, different objects. Ask: *What can you tell me about the size of this object? Is it big or small?* Emphasise the words 'size', 'big' and 'small'.

Teaching

- Hold up two similar objects of different sizes, e.g. two building blocks. Say: *This block is bigger than this block.* Emphasise the word 'bigger'.
- Then hold up two small-world resources of different sizes, e.g. bears. Say: *This bear is smaller than this bear.* Emphasise the word 'smaller'.
- Repeat for other pairs of similar objects, e.g. two recycled boxes, two shoes or two soft toys.
- Hold up two *different* objects of different sizes, e.g. a book and a crayon. Say: *This book is bigger than this crayon. The crayon is smaller than the book.*

- Repeat for other pairs of different objects. Include examples of big objects that are light (e.g. a balloon), and small objects that are heavy (e.g. a stapler). Ask children to tell you about the size of the two objects, using the words 'bigger' and 'smaller'.
- Hold up, or point to, an object, e.g. a shoe. Ask individual children to go and point to something that is bigger than/smaller than your object.
- Show pages 24 and 25 (The farmyard) of the *Reading Anthology*. Discuss the picture. Ask children to describe what they can see, using the words 'big', 'bigger', 'small' and 'smaller'.

Explore in groups

Sorting by size

- Provide two sorting hoops and a selection of objects suitable for sorting by size (but that focus on mass). Invite children to sort the objects into two sets: 'big' and 'small'. Ask them to tell you how they sorted the objects.

Smaller hunt

- Provide bags, and a variety of objects (e.g. soft toy, building block, small recycled box, sock). Invite children to take an object and a bag, and go on a size hunt around the room (and outdoors, if possible). Ask them to collect objects that are smaller than their object.

Variation: ask children to look for objects that are bigger than their object.

Maths Foundation Reading Anthology
(adult-led)

- Together, look at pages 24 and 25 (The farmyard) of the *Reading Anthology*. Repeat the last part of the *Teaching* activity with smaller groups of children.

Session 2: Describe masses

Children begin to talk about the mass of objects. They use the words 'heavy' and 'light'.

You will need:

selection of objects suitable for describing mass, e.g. feathers, balloons, scraps of paper, paper clips, pencils, crayons, interlocking cubes, beads, balls, books, shoes, building blocks, bean

Teaching notes

bags, small-world resources, soft toys, Home corner props, recycled boxes and packaging, natural materials; sorting hoops; small pots or cups; outdoor equipment and resources

Getting started

- Before the session, place on a table a selection of objects suitable for describing mass.
- Allow children to spend a few minutes picking up and feeling the objects.

Teaching

- Hold up a small, light object (e.g. an interlocking cube) and a bigger, heavier object (e.g. a large stone). Say: *This stone is heavy and this cube is light.* Emphasise the words 'heavy' and 'light'.
- Pass the two objects around the class, allowing each child to hold them. Ask children how they feel, reinforcing the words 'heavy' and 'light'.
- Point to, or hold up, different objects. Say, for example: *This is a heavy box. This pencil is light. This shoe is heavy. This paper clip is light.* Each time, give the object to different children to hold (if possible).
- Try to include examples of big objects that are light, and small objects that are heavy. This will help to address the common misconception that bigger objects are always heavier.
- Give an object to a child to hold. Ask, for example: *What can you tell me about the mass of this object?* Emphasise the word 'mass'. Explain: *Mass tells us how light or heavy something is.* Ask the child to pass the object around to the others. Repeat for other objects.
- Ask individual children to go and find something that is heavy/light.

Explore in groups

Sorting by mass

- Provide two sorting hoops and a selection of objects suitable for sorting by mass. Invite children to sort the objects into two sets: 'heavy' and 'light'. Ask them to tell you how they sorted the objects.

Outdoor play

- Let children play freely, using outdoor equipment and resources. Occasionally ask them to tell you about the objects they are playing with: *Does this feel heavy or light?*

Heavy and light pots

- Give each child two small pots or cups. Ask them to fill one pot with objects from around the room to make the pot as heavy as possible. They fill the other pot with objects that make it as light as possible.

Session 3: Compare masses

Children compare the masses of two objects. They use the words 'heavier', 'lighter' and 'the same'.

You will need:

Maths Foundation Reading Anthology; selection of different objects suitable for describing and comparing masses of two objects (include some objects that are the same mass), e.g. feathers, balloons, scraps of paper, paper clips, pencils, crayons, interlocking cubes, beads, balls, books, shoes, building blocks, bean bags, small-world resources, soft toys, Home corner props, recycled boxes and packaging, natural materials; playdough; *Maths Foundation Activity Book C*; pencils

Getting started

- Before the session, place on a table a selection of objects suitable for comparing mass. Also have two balls of playdough of distinctly different sizes.
- Hold up a heavy object and a light object: *This ... is heavy. This ... is light.* Emphasise the words 'heavy' and 'light'.
- Pass the objects around so that all children get to hold them.
- Repeat for other objects. Include examples of big objects that are light, and small objects that are heavy.

Teaching

- Hold up a pair of objects, e.g. a bean bag and a feather. Hold one on each hand and exaggerate tipping a little towards the side with the heavier object. Say: *This bean bag is heavier than this feather. The feather is lighter than the bean bag.* Emphasise the words 'heavier' and 'lighter'.

- Repeat for another pair of objects.
- Ask children to stand up. Pass two objects to a child, e.g. a leaf and a large stone. Ask them to say which one feels heavier. Pass the objects round to the rest of the children so they can feel the difference themselves.
- Repeat for other pairs of objects. Once again, include big objects that are light, and small objects that are heavy. Also include objects that have the same mass, e.g. *These two shoes have the same mass/weigh the same.* Emphasise the phrase: 'the same'.
- Hold up two different-sized balls of playdough. Ask: *Which do you think will be heavier?* Pass the two balls around for all children to hold. *So, which is the heavier ball?*
- Take the heavier ball and roll it out into a worm. Ask: *Do you think that this is still the heavier piece of playdough?* Discuss how objects don't change their mass just by changing their shape: *conservation.*
- Pass an object to a child so they can feel its mass. Ask them to go and find something that is heavier/lighter than your object. Repeat with different children and objects.
- Show pages 24 and 25 (The farmyard) of the *Reading Anthology*. Discuss the picture. Ask children to describe and compare what they can see in the picture, using the words 'heavy', 'heavier', 'light' and 'lighter'.

Explore in groups

Which is heavier?

- Provide a selection of objects suitable for comparing masses. Invite children to work in pairs. Child A closes their eyes and holds out their hands. Child B chooses two objects and places one in each of Child A's hands. Child A then decides which of the objects is heavier. Both children then discuss and agree whether Child A is correct.

Variation: Child A decides which of the two objects is lighter.

Heavier and lighter

- Provide a selection of objects suitable for comparing masses. Invite children to work in pairs. Each child chooses an object and gives it to their partner. Each child then chooses two other objects: one heavier and one lighter than the object they were given. Children then check each other's objects. Encourage

children to use the words 'heavier', 'lighter' and 'the same'.

Maths Foundation Reading Anthology (adult-led)

- Together, look at pages 24 and 25 (The farmyard) of the *Reading Anthology*. Repeat the last part of the *Teaching* activity with smaller groups of children.

Maths Foundation Activity Book C (adult-led)

Page 15 – Heavier

Session 4: Balance scales

Children use balance scales to compare masses.

You will need:

two identical buckets (or bags); balance scales; selection of objects suitable for describing and comparing masses of two objects (include some objects that are the same mass), e.g. feathers, balloons, scraps of paper, paper clips, pencils, crayons, interlocking cubes, beads, balls, books, shoes, building blocks, bean bags, small-world resources, soft toys, Home corner props, recycled boxes and packaging, natural materials; playdough; *Maths Foundation Activity Book C*; pencils

Getting started

- Before the session, place a heavy object in one bucket and a light object in the other bucket. Also, on a table place a balance scale and a selection of objects suitable for describing and comparing mass. Only include objects that will fit into the buckets and on the balance scale.
- Remind children of the words 'heavy', 'heavier', 'light', 'lighter' and 'the same'. Pick up a heavy object and a light object. Say: *This … is heavy. This … is light.*
- Pass the objects around so all the children get to hold them.
- Then say: *The … is heavier than the … The … is lighter than the …*
- Repeat several times, including some big objects that are light, and small objects that are heavy.

Teaching notes

Teaching

- Pick up the two buckets, one in each hand. Acting as a 'human balance scale', exaggerate tilting towards the heavier side, as if that arm is being pulled downwards. Ask: *Which of these two buckets do you think has the heavier object in it?* Discuss responses. Then reveal the contents of each bucket. Confirm that the bucket with the heavier object had the stronger downward pull: *The bucket that had the … was closer to the ground as it was heavier for me to lift. The bucket with the … I could hold up easier, as it was lighter.* Emphasise the words 'heavier' and 'lighter'.
- Put a different pair of objects in the buckets. Give the buckets to individual children to hold, and ask them to say which is heavier and which is lighter. Repeat several times.
- Point to the balance scale. Say: *This is called a balance scale. We can use this to help us see which object is heavier.*
- Choose two objects. Pass them to the children, so they all get a chance to hold them. Ask: *Which do you think is heavier, the … or the …?*
- Say: *We can use the balance scale to check.* Place the two objects on either side of the scale. Ask: *What's happened?* Discuss responses, highlighting that the side with the heavier object goes down, and the side with the lighter object goes up. Say: *You were right: the … is heavier than the … The … is lighter than the …*
- Repeat for several pairs of objects. Include some pairs where the bigger object is lighter, and some pairs that have the same mass. Each time, ask a different child to place the objects on the balance scale to check.

Explore in groups

Balance scales exploration

- Provide balance scales and a selection of objects suitable for comparing masses. (Only include objects that will fit on a balance scale.) Let children play and explore freely using the balance scales. Occasionally, ask children to tell you which object is heavier and which is lighter, and how they know.

Heavier objects

- Provide several balance scales and a selection of objects suitable for comparing masses.

(Only include objects that will fit on a balance scale.) Place an object on one side of each balance scale (a different object for each scale). Invite children to find objects that are heavier than the ones already on the scales.

Variation: ask children to find objects that are lighter.

Playdough station

- Provide playdough and balance scales. Challenge children to make two playdough balls that balance. When they have done this, ask them to remove the balls from the balance scales and change the shape of each ball. Ask: *Do the two pieces of playdough still balance?*

Maths Foundation Activity Book C (adult-led)

Pages 16 and 17 – Balance scales

Session 5: Describe and compare capacities (A)

Children begin to describe and compare the capacities of two containers. They use the phrases: 'holds more', 'holds less' and 'holds the same'.

You will need:

selection of containers suitable for describing and comparing capacities (including some transparent containers and some identical containers), e.g. buckets, baskets, jugs, pots and pans, mugs, cups, glasses, measuring cups, ladles, spoons, measuring spoons, food storage containers, recycled packaging such as tins, cartons, plastic bottles; selection of resources suitable for filling containers, e.g. large jug of water, sand, dried lentils, pebbles, shells, marbles, centimetre cubes; water tray; sand tray; large empty tray; *Maths Foundation Reading Anthology*

Getting started

- Before the session, place on a table a selection of containers suitable for describing and comparing capacity, and a selection of resources suitable for filling containers. Ensure that the water tray, sand tray and/or a large empty tray are nearby.

- Allow children to spend a few minutes looking at, and picking up, the containers and resources.

Teaching

- Begin to introduce children to the term capacity. Explain: *When we talk about how much a container can hold, we are talking about the capacity of the container. Capacity tells us the most that a container can hold.* Emphasise the word 'capacity'.
- Ask a child to come to the table, choose a container, and show it to the other children. Ask: *What can you tell me about this container? Is it tall or is it short? Is it wide or is it narrow? Do you think it could hold a lot of water/sand/shells or only a little?* Repeat for a few other children and containers.
- Next hold up two different containers. Ask: *Which of these containers do you think holds more? Why? Which of these containers do you think holds less? How do you know?* Emphasise the phrases 'holds more' and 'holds less'.
- Discuss responses. Guide children to consider all of the containers' dimensions.
- Say: *We are going to find out which container holds more by filling each container with water/sand.* Fill the smaller container. Then pour its contents into the larger container. Point to the larger container and say: *I could put more water/sand into this container. So, this container holds more than this container.* Again, emphasise the phrase 'holds more'.
- Repeat several times, for different pairs of containers.
- Repeat again, but this time fill the larger container first. Then stand over the water/sand tray, and pour its contents into the smaller container. Continue to pour until the smaller container overflows. Point to the smaller container and say: *I can't put any more water/sand into this container. So, this container holds less than this container.* Again, emphasise the phrase 'holds less'.
- Repeat several times, for different pairs of containers. If you have only used water for filling so far, switch to a different resource.
- End by comparing the capacities of two identical containers in the same way. Introduce children to the phrase 'holds the same'.

Explore in groups

Capacity exploration

- Provide a selection of containers suitable for comparing capacities, and a selection of resources suitable for filling containers. Let children play and explore freely. Ask children to tell you what they have discovered.

Comparing capacities

- Provide a selection of containers suitable for comparing capacities, and a selection of resources suitable for filling containers. Invite children to compare the capacities of pairs of containers by filling one and pouring the contents into the other. Ask questions such as: *Which container holds more? Have you found any containers that hold the same?*

Maths Foundation Reading Anthology
(adult-led)

- Together, look at pages 24 and 25 (The farmyard) of the *Reading Anthology*. Briefly discuss the picture. Ask children to describe and compare what they can see in the picture, using the phrases 'holds more', 'holds less' and 'holds the same'.

Session 6: Describe and compare capacities (B)

Children continue to describe and compare the capacities of two containers.

You will need:

selection of containers suitable for describing and comparing capacities (including some transparent containers, some identical containers and some different-shaped containers with the same capacity), e.g. buckets, baskets, jugs, pots and pans, mugs, cups, glasses, measuring cups, ladles, spoons, measuring spoons, food storage containers, recycled packaging such as tins, cartons, plastic bottles; selection of resources suitable for filling containers, e.g. large jug of water, sand, dried lentils, pebbles, shells, marbles, centimetre cubes; water tray; sand tray; large empty tray; *Maths Foundation Activity Book C*; pencils

Teaching notes

Getting started

- Before the session, place on a table a selection of containers suitable for describing and comparing capacity, and a selection of resources suitable for filling containers. Ensure that a water tray, sand tray and/or a large empty tray is nearby.
- Hold up a container. Say: *Remember, when we talk about how much a container can hold, we are talking about the capacity of the container. The word capacity means the most that a container can hold.* Emphasise the word 'capacity'.

Teaching

- Hold up another container and ask: *Who can find a container that they think holds more than this one?* Emphasise the phrase 'holds more'. Choose a child to come and find a container. Ask: *Arjun, why do you think this container holds more than my container?* Discuss the child's response. Guide them to consider all of the containers' dimensions.
- Ask another child to fill your container, e.g. with water or sand. Then pour its contents into the container that the first child chose.
- Point to the child's container and say: *We could put more water/sand into this container. So, Arjun's container holds more than mine.*
- Repeat, choosing a different container, and asking a child to find a container that holds less than yours. Emphasise the phrase 'holds less'. Ask another child to help you check, but this time fill your container first. Stand over the water/sand/empty tray and pour the contents into the container chosen by the child.
- Point to the child's container and say: *We couldn't put all the water/sand that was in my container into Faith's container. So, Faith's container holds less than mine.*
- Repeat several times, for different containers. If you have only used water for filling so far, switch to a different resource.
- End by comparing two different-shaped containers with the same capacity (e.g. a half-litre plastic bottle and a half-litre ice cream tub). Say: *These containers look different, but they both hold the same amount.*

Explore in groups

Holds more

- Provide a selection of containers suitable for comparing capacities, and a selection of resources suitable for filling containers. Invite children to work in pairs. Each child chooses a container and gives it to their partner. Each child then chooses another container that they think holds more than the container they were given. Encourage children to check by filling one and pouring the contents into the other.

Variation: ask children to choose a container that holds less than the container they were given.

Holds the same

- Provide a selection of containers suitable for comparing capacities, and a selection of resources suitable for filling containers. Challenge children to look for two containers that they think have the same capacity. Encourage them to check by filling one and pouring the contents into the other.

Maths Foundation Activity Book C (adult-led)

Page 18 – Holds more

Session 7: Describe volume

Children begin to describe volume. They use the phrases 'full', 'nearly full', 'empty' and 'nearly empty'.

You will need:

Maths Foundation Reading Anthology; selection of containers suitable for describing volume (including some transparent containers, some identical containers, and some different-shaped containers with the same capacity), e.g. buckets, baskets, jugs, pots and pans, mugs, cups, glasses, measuring cups, ladles, spoons, measuring spoons, food storage containers, recycled packaging such as tins, cartons, plastic bottles; selection of resources suitable for filling containers, e.g. large jug of water, sand, dried lentils, pebbles, shells, marbles, centimetre cubes; water tray; sand tray; large empty tray; PCM 27; *Maths Foundation Activity Book C*; coloured pencils and crayons

Getting started

- Before the session, place on a table a selection of containers suitable for describing volume, and a selection of resources suitable for filling containers. Ensure that a water tray, sand tray and/or a large empty tray is nearby.
- Hold up two containers with distinctly different capacities, e.g. a pot and a mug. Remind children of the term capacity. Say: *Remember, the word capacity means the most that a container can hold.*

Teaching

- Hold up another container, e.g. a jar. Say: *This jar has nothing in it. There is nothing at all inside. It is empty.* Emphasise the word 'empty'.
- Ask a child to hold the jar over the water tray. Slowly pour water into the jar until it is full. Continue pouring until the jar overflows. Ask: *What has just happened?* Discuss responses and introduce the word 'full': *The jar is now full of water.*
- Pour the water back into the jug. Say: *Now the jar is empty again.*
- Choose another child to hold the jar. Once again slowly pour water from into the jar until it is nearly full. Ask: *Now what can you tell me about the water in the jar?* Discuss responses and introduce the phrase 'nearly full': *Now the jar is nearly full of water.*
- Pour most of the water back into the jug. Ask: *Now what can you tell me about the water in the jar?* Discuss responses and introduce the phrase 'nearly empty': *Now the jar is nearly empty.*
- Repeat several times, using different containers and filling them with different resources.
- Show pages 24 and 25 (The farmyard) of the *Reading Anthology*. Briefly remind children of the picture. Ask them to describe and compare what they can see in the picture using the phrases 'full', 'nearly full', 'empty' and 'nearly empty'.

Explore in groups

Filling containers

- Provide a selection of containers suitable for describing volume, and a selection of resources suitable for filling containers. Encourage children to choose four containers

and fill them so that one is full, one is nearly full, one is nearly empty and one is empty.

Filling glasses

- Provide the pictures of glasses from PCM 27. Invite children to colour the four glasses to show that one is full, one is nearly full, one is nearly empty and one is empty.

Variation: give children the four bowls from PCM 27 instead, and ask children to draw e.g. breakfast cereal, with one bowl full, one nearly full, one nearly empty and one empty.

Maths Foundation Activity Book C (adult-led)

Page 19 – Full and empty

When children have completed the page, ask them to point to the objects they have not circled and say whether they are nearly full or nearly empty.

Session 8: Compare volumes

Children compare the volumes of two containers. They use phrases such as 'has more', 'has less' and 'has the same'.

You will need:

pairs of identical, transparent containers suitable for describing and comparing volume, e.g. jugs, glasses, food storage containers, recycled packaging such as bottles; selection of resources suitable for filling containers, e.g. large jug of water, sand, dried lentils, pebbles, shells, marbles, centimetre cubes; PCM 27; coloured pencils and crayons

Getting started

- Before the session, place on a table pairs of identical, transparent containers, and a selection of resources suitable for filling containers.
- Hold up two identical containers, e.g. two glasses. Ask: *What can you tell me about these two glasses? That's right, they're exactly the same size.*
- Pour different amounts of water into each glass. Ask: *What can you tell me about the water in these two glasses? Which has more water? Which has less water? How do you know?* Emphasise the phrases 'has more' and

Teaching notes

'has less'. Discuss responses, encouraging children to explain their reasoning.

- Then point to each glass in turn and say: *This glass has more water than this glass. And this glass has less water than this glass.*

Teaching

- Repeat the *Getting started* activity several times, with different pairs of identical containers. Fill some pairs of containers with solid materials instead of water.
- Show another pair of identical containers. Partly fill one of the containers, e.g. with dried lentils. Hold up the other container. Ask: *Who can come and put lentils into this container so that there are more lentils in it than there are in this one?*
- Once the child has done this, ask: *Did Samuel put more lentils into this container?* Then say: *So, Samuel's container has more lentils in it than my container. My container has less lentils in it than Samuel's container.* Emphasise the phrases 'has more' and 'has less'.
- Repeat several times, using different pairs of identical containers and different resources to fill them. Include some examples that involve children trying to match the amount you put in your container. Emphasise the phrase 'has the same'.

Explore in groups

Volume exploration

- Provide pairs of identical, transparent containers, and a selection of resources suitable for filling containers. Let children play and explore freely. Occasionally, ask children to tell you which of two containers has more and which has less, and how they know.

Matching volumes

- Provide pairs of identical, transparent containers, and a selection of resources suitable for filling containers. Encourage children to play in pairs. They need an identical container each. They take turns to part-fill their container, and challenge their partner to fill their container to match. Ask them to explain how they know that their containers now have the same.

More and less water

- Provide the pictures of glasses from PCM 27. Invite children to work in pairs. Each child colours their glasses to show four different quantities of water. Children then take turns to point to two of their glasses and ask their partner to say which glass has more (or less) water.

Variation: give children the four bowls from PCM 27 instead, and ask children to draw different quantities of e.g. peas.

Assessment opportunities

Assess children's learning against the objectives for this unit, using the guidance on formative assessment on pages 24–25, and record your observations in the Unit 14 progress tracking grid on page 33. The relevant pages of *Activity Book C* can also be used for assessment.

Can the children:

- describe and compare the sizes of objects, using the words: 'big', 'bigger', 'small' and 'smaller'?
- describe and compare the masses of objects, using the words: 'heavy', 'heavier', 'light', 'lighter' and 'the same'?
- compare the capacities of containers, using the phrases: 'holds more', 'holds less' and 'holds the same'?
- describe and compare volumes in containers, using the phrases: 'full', 'nearly full', 'nearly empty', 'empty', 'has more', 'has less' and 'has the same'?

Unit 15 3D shapes

Theme: Technology

Overview

In this unit, children explore shapes in the world around them. They are introduced to the 3D ('solid') shapes: cube, cuboid, cylinder and sphere. They learn to name 3D shapes and recognise them in different sizes and orientations. They compare and sort 3D shapes and begin to use simple shape vocabulary to describe them.

During this unit, find opportunities throughout the day to use the *Number fluency* activities on pages 198–207, continue to practise counting and recognising, reading, writing and comparing numbers to 10, as well as to develop children's understanding of addition and subtraction.

Learning objectives

Geometry and Measure – Understanding shape	• **6b Identify, describe, compare and sort 3D shapes.**
Statistics	• *10a Sort, represent and describe data using concrete materials or pictorial representations.*

Learning objectives in italics have been taught previously. In this unit they are consolidated and/or extended.

Learning objectives in bold are taught for the first time in this unit.

Vocabulary

3D (3-dimensional), shape, curved, flat, round, solid, corner, face, surface, end, sort, cube, cuboid, cylinder, sphere, circle, square, rectangle, slide, roll, stack

Making connections

English: Using technology

Science: Changing materials

Preparation

You will need:

- *Maths Foundation Reading Anthology*, pages 2–5, 12 and 13, 18 and 19.
- *Maths Foundation Activity Book C*, pages 20–23.
- Digital Tool: Shape set.
- PCMs 25 and 26.
- Magazines, comics, catalogues, toy packaging, etc.
- Several real-world examples of cubes, cuboids, cylinders and spheres.
- 3D geometric shapes.

- Sorting hoops.
- Labels: 'cubes', 'not cubes', 'cuboids', 'not cuboids', 'cylinders', 'not cylinders', 'spheres' and 'not spheres'.
- Trays.
- Piece of fabric.
- Non-transparent ('feely') bags.
- Large books.
- Painting and drawing station: paper, pencils, coloured pencils or crayons, paint, water, paintbrushes, aprons.

Teaching notes

- Playdough station: playdough, 3D geometric shapes.
- Sand station: sand tray/pit: 3D geometric shapes, dishes.
- Construction station: construction materials, e.g. building blocks and recycled boxes and packaging, especially cubes, cuboids, cylinders and spheres, glue or tape, shape cards from PCMs 25 and 26.

Before starting

As part of Session 1, start to create a wall and table display of 3D shapes. Throughout the unit ask children to look for examples of cubes, cuboids, cylinders and spheres in magazines, comics, catalogues, etc. as well as real-world examples to include in the display.

Children will get the most out of this unit if they already have some experience of sorting familiar objects using criteria such as size and colour. They should be able to use everyday words to explain choices, including 'not'.

Children also need to be familiar with the language of movement, in particular the words: 'slide' and 'roll' which were introduced in Unit 3.

Before teaching this unit, look back at the Assessment section at the end of Unit 5 (page 76), to identify the prerequisite learning for this unit.

Common difficulties

Some children may have difficulty in understanding the difference between 2D and 3D shapes (for example, thinking that a circle and a sphere are the same). Remind children regularly that 2D shapes are flat and 3D shapes are solid. Use both real-world and geometric 2D and 3D shapes to reinforce this.

Children may also be confused about representations of 3D shapes in pictures. Help children distinguish between a representation of a 3D shape and an actual 3D shape by saying, for example: *Here is a basketball. It is a sphere. This is a picture of basketball. A basketball is a sphere.*

Session 1: Recognise 3D shapes in everyday life

Children begin to recognise and name 3D shapes in everyday life.

You will need:

real-world examples of cubes, cuboids, cylinders and spheres; sorting trays; 3D geometric shapes; paper; coloured pencils or crayons, or paint, water, paintbrushes, aprons; *Maths Foundation Reading Anthology*

Getting started

- Begin by pointing out examples of different real-world 3D shapes within the classroom. Point to different shapes in turn, saying the name of the shape. Focus on cubes, cuboids, cylinders and spheres. Include a selection of shapes of different sizes and from different perspectives.
- It is important to make it clear to children that you are referring to the entire 3D shape. Where possible, hold up the object and show it to the children from different perspectives. Point to the shape's faces, corners and edges to emphasise the three-dimensional form.
- Occasionally, ask children to describe a shape using their own words. (At this stage, do not introduce children to the *properties* of cubes, cuboids, cylinders and spheres.) Ask them to say what is the same and what is different about various 3D shapes. Encourage them to recognise and name the 3D shapes and describe them as '3D' or 'solid' shapes.

Teaching

- Take children on a 3D shape hunt around the classroom and outdoors.
- Ask children to point to and name different 3D shapes. Where appropriate, encourage them to touch the faces/surfaces, edges and corners of different shapes.

- Once children begin to become familiar with cubes, cuboids, cylinders and spheres, arrange them into four groups: a cube group, a cuboid group, a cylinder group and a sphere group. Ask all the members of each group to find, or point to, an example of their group's shape. Reinforce good examples of each shape, e.g. *Yes, Marco, that's a cube. We can say that a cube is a solid shape. We can also say that a cube is a 3D shape.*
- Repeat several times, changing each group's shape.

Explore in groups

Painting and drawing station

- Invite children to draw or paint a picture based on their 3D shape hunt. Ask children to name some of the shapes in their pictures.

Sorting

- Provide sorting trays, 3D geometric shapes and examples of different real-world 3D shapes. Encourage children to sort the resources using their own criteria. Ask them to explain how they have sorted. If they have not used shape as criteria, ask them to consider sorting by shape. Can they name any of the 3D shapes?

Maths Foundation Reading Anthology (adult-led)

- Together, look at different pages in the *Reading Anthology*. Discuss each picture with the children. Ask them to point to and name examples of cubes, cuboids, cylinders and spheres. For example: pages 2 and 3 (At home): present (cube), cake and saucepan (cylinder), balls of wool (sphere); pages 4 and 5 (In the garden): buckets (cylinder), balls, nuts in the tree (spheres); pages 12 and 13 (The fruit stall): crates/boxes (cuboids), apples, cherries and plums (spheres).

Session 2: Recognise and name cubes

Children begin to recognise and name cubes, including different sizes and perspectives, and real-world examples.

You will need:

3D geometric shapes; real-world examples of cubes and other 3D shapes (cuboids, cylinders and spheres); two sorting hoops; labels: 'cubes' and 'not cubes'; tray; piece of fabric; *Maths Foundation Reading Anthology*; paper; coloured pencils or crayons; PCMs 25 and 26

Getting started

- Show children a 3D geometric cube. Briefly discuss its properties: a 3D, solid shape with 6 square faces. At this stage, do not expect children to recall these properties. They should just be able to recognise and name cubes, and distinguish between shapes that are cubes and shapes that are *not* cubes, especially cuboids.
- Rotate the cube to show it from different perspectives.
- Demonstrate how a cube slides when pushed along any of its faces. Emphasise the word 'slide'.
- Show other 3D geometric cubes of different sizes and several real-world examples of cubes. Constantly reinforce the words 'cube', '3D' and 'solid'.

Teaching

- Show children a shape (3D geometric shape or real-world example) that is *not* a cube, such as a sphere. Ask: *Is this shape a cube? What's the same about this shape and a cube? What's different about them?*
- Repeat for other 3D shapes that are *not* cubes.
- Place two sorting hoops on the floor. Label them: 'cubes' and 'not cubes'. Ask a child to come and choose a shape (either a 3D geometric shape or a real-world example). Ask: *Is this shape a cube?* Ask the child to place the shape into the correct hoop.

Teaching notes

- Repeat several times.
- Ask children to point to any other examples of cubes they can see, inside the classroom and outdoors.

Explore in groups

Take the cubes

- Place 3D geometric shapes and real-world examples of cubes, cuboids, cylinders and spheres on a tray. (Include more cubes than any other shape.) Cover it with a piece of fabric. Invite children to work in pairs or groups. They take turns to feel for a shape which is a cube, take it from the tray and place it beside them. When the tray is empty the winner is the child who has the most cubes.

Maths Foundation Reading Anthology (adult-led)

- Together, look at pages 18 and 19 (At the supermarket) of the *Reading Anthology*. Discuss the examples of cubes in the picture. Ask children to draw some things that are cube-shaped. They can use ideas from the *Reading Anthology*, or their own ideas.

Building with shapes

- Provide 3D geometric shapes and real-world examples of 3D shapes (cubes, cuboids, cylinders and spheres). Let children build models with them. You could encourage children to build something linked to the theme 'technology', e.g. a robot. Ask children to tell you about the shapes they have used. Ask: *Were some shapes harder to build with? Why do you think that is?*

Cubes and not cubes

- Provide a set of cube, cuboid, cylinder and sphere cards from PCMs 25 and 26. Invite children to sort the cards into two sets: 'cubes' and 'not cubes'.

Variation: use 3D geometric shapes instead of the cards.

Session 3: Recognise and name cuboids

Children begin to recognise and name cuboids, including different sizes and perspectives, and real-world examples.

You will need:

3D geometric shapes; real-world examples of cuboids and other 3D shapes (cubes, cylinders and spheres); two sorting hoops; labels: 'cuboids' and 'not cuboids'; tray; piece of fabric; *Maths Foundation Reading Anthology*; paper; coloured pencils or crayons; PCMs 25 and 26; *Maths Foundation Activity Book C*

Getting started

- Show children a 3D geometric cuboid. Briefly discuss its properties: a 3D, solid shape with 6 faces that are rectangles. At this stage do not expect children to recall these properties. They should just be able to recognise and name cuboids, and distinguish between shapes that are cuboids and shapes that are *not* cuboids, especially cubes.
- Rotate the cuboid to show it from different perspectives.
- Demonstrate how a cuboid slides when pushed along any of its faces. Emphasise the word 'slide'.
- Show other 3D geometric cuboids of different sizes and several real-world examples of cuboids. Constantly reinforce the words 'cuboid', '3D' and 'solid'.

Teaching

- Show children a shape (3D geometric shape or real-world example) that is *not* a cuboid, such as a cylinder. Ask: *Is this shape a cuboid? What's the same about this shape and a cuboid? What's different about them?*
- Repeat for other 3D shapes that are *not* cuboids.
- Place two sorting hoops on the floor. Label them 'cuboids' and 'not cuboids'. Ask a child to come and choose a shape (either a 3D geometric shape or a real-world example). Ask: *Is this shape a cuboid?* Ask the child to place the shape into the correct hoop.
- Repeat several times.

- Ask children to point to any other examples of cuboids they can see, inside the classroom and outdoors.

Explore in groups

Take the cuboids

- Place 3D geometric shapes and real-world examples of cubes, cuboids, cylinders and spheres on a tray (include more cuboids than any other shape). Cover it with a piece of fabric. Invite children to work in pairs or groups. They take turns to feel for a shape which is a cuboid, take it from the tray and place it beside them. When the tray is empty the winner is the child who has the most cuboids.

Maths Foundation Reading Anthology (adult-led)

- Together, look at pages 18 and 19 (At the supermarket) of the *Reading Anthology*. Discuss the examples of cuboids in the picture. Ask children to draw some things that are cuboid-shaped. They can use ideas from the *Reading Anthology*, or their own ideas.

Cuboids and not cuboids

- Provide a set of cube, cuboid, cylinder and sphere cards from PCMs 25 and 26. Invite children to sort the cards into two sets: 'cuboids' and 'not cuboids'.

Variation: use 3D geometric shapes instead of the cards.

Maths Foundation Activity Book C (adult-led)

Page 20 – Cubes and cuboids

Session 4: Recognise and name cylinders

Children begin to recognise and name cylinders, including different sizes and perspectives, and real-world examples.

You will need:

3D geometric shapes; real-world examples of cylinders and other 3D shapes (cubes, cuboids and spheres); two sorting hoops; labels:

'cylinders' and 'not cylinders'; tray; piece of fabric; *Maths Foundation Reading Anthology*; paper; coloured pencils or crayons; playdough; PCMs 25 and 26

Getting started

- Show children a 3D geometric cylinder. Briefly discuss its properties: a 3D, solid shape with 2 end faces that are circles joined together by 1 curved face (or surface). At this stage, do not expect children to recall these properties. They should just be able to recognise and name cylinders, and distinguish between shapes that are cylinders and shapes that are *not* cylinders.
- Rotate the cylinder to show it from different perspectives.
- Demonstrate how a cylinder slides when pushed along either of its two circular ends. Then show how it rolls along its curved face/surface. Emphasise the words 'slide' and 'roll'.
- Show other 3D geometric cylinders of different sizes and several real-world examples of cylinders. Constantly reinforce the words 'cylinder', '3D' and 'solid'.

Teaching

- Show children a shape (3D geometric shape or real-world example) that is *not* a cylinder, such as a sphere. Ask: *Is this shape a cylinder? What's the same about this shape and a cylinder? What's different about them?*
- Repeat for other 3D shapes that are *not* cylinders.
- Place two sorting hoops on the floor. Label them 'cylinders' and 'not cylinders'. Ask a child to come and choose a shape (either a 3D geometric shape or a real-world example). Ask: *Is this shape a cylinder?* Ask the child to place the shape into the correct hoop.
- Repeat several times.
- Ask children to point to any other examples of cylinders they can see, inside the classroom and outdoors.

Teaching notes

Explore in groups

Take the cylinders

- Place 3D geometric shapes and real-world examples of cubes, cuboids, cylinders and spheres on a tray (include more cylinders than any other shape). Cover it with a piece of fabric. Invite children to work in pairs or groups. They take turns to feel for a shape which is a cylinder, take it from the tray and place it beside them. When the tray is empty the winner is the child who has the most cylinders.

Maths Foundation Reading Anthology (adult-led)

- Together, look at pages 18 and 19 (At the supermarket) of the *Reading Anthology*. Discuss the examples of cylinders in the picture. Ask children to draw some things that are cylinder-shaped. They can use ideas from the *Reading Anthology*, or their own ideas.

Playdough station

- Provide a selection of 3D geometric shapes, including several cylinders. Let children explore pressing the shapes into playdough to see what shape each face can make. Ask them to tell you what they have found out.

Cylinders and not cylinders

- Provide a set of cube, cuboid, cylinder and sphere cards from PCMs 25 and 26. Invite children to sort the cards into two sets: 'cylinders' and 'not cylinders'.

Variation: use 3D geometric shapes instead of the cards.

Session 5: Recognise and name spheres

Children begin to recognise and name spheres, including different sizes and perspectives, and real-world examples.

You will need:

3D geometric shapes; real-world examples of spheres and other 3D shapes (cubes, cuboids and cylinders); two sorting hoops; labels: 'spheres' and 'not spheres'; tray; piece of fabric; *Maths Foundation Reading Anthology*; paper; coloured pencils or crayons; PCMs 25 and 26; *Maths Foundation Activity Book C*

Getting started

- Show children a 3D geometric sphere. Briefly discuss its properties: a 3D, solid shape that is round, like a ball, with 1 face (or surface). At this stage, do not expect children to recall these properties. They should just be able to recognise and name spheres, and distinguish between shapes that are spheres and shapes that are *not* spheres.
- Rotate the sphere to show that it is identical from any perspective.
- Demonstrate how a sphere rolls along its curved face/surface. Emphasise the word 'roll'.
- Show other 3D geometric spheres of different sizes and several real-world examples of spheres. Constantly reinforce the words 'sphere', '3D' and 'solid'.

Teaching

- Show children a shape (3D geometric shape or real-world example) that is *not* a sphere, such as a cube. Ask: *Is this shape a sphere? What's the same about this shape and a sphere? What's different about them?*
- Repeat for other 3D shapes that are *not* spheres.
- Place two sorting hoops on the floor. Label them 'spheres' and 'not spheres'. Ask a child to come and choose a shape (either a 3D geometric shape or a real-world example). Ask: *Is this shape a sphere?* Ask the child to place the shape into the correct hoop.
- Repeat several times.
- Ask children to point to any other examples of spheres they can see, inside the classroom and outdoors.

Explore in groups

Take the spheres

- Place 3D geometric shapes and real-world examples of cubes, cuboids, cylinders and spheres on a tray (include more spheres than any other shape). Cover it with a piece of fabric. Invite children to work in pairs or groups. They take turns to feel for a shape which is a sphere, take it from the tray and place it beside them. When the tray is empty the winner is the child who has the most spheres.

Maths Foundation Reading Anthology
(adult-led)

- Together, look at pages 18 and 19 (At the supermarket) of the *Reading Anthology*. Discuss the examples of spheres in the picture. Ask children to draw some things that are sphere-shaped. They can use ideas from the *Reading Anthology*, or their own ideas.

Spheres and not spheres

- Provide a set of cube, cuboid, cylinder and sphere cards from PCMs 25 and 26. Invite children to sort the cards into two sets: 'spheres' and 'not spheres'.

Variation: use 3D geometric shapes instead of the cards.

Maths Foundation Activity Book C (adult-led)

Page 21 – Cylinders and spheres

Session 6: Compare 3D shapes

Children compare two 3D shapes. They say what is the same and what is different.

You will need:

Maths Foundation Reading Anthology; Shape set Digital Tool; PCMs 25 and 26; 3D geometric shapes; non-transparent ('feely') bags; real-world examples of cubes, cuboids, cylinders and spheres

Getting started

- Begin by holding up shape cards from PCMs 25 and 26. Ask children to name each shape.

Teaching

- Show pages 18 and 19 (At the supermarket) of the *Reading Anthology*. Ask individual children to come and point to a sphere, e.g. *Nina, can you point to something that is the shape of a sphere? What shape are the soda cans/oranges …?*
- Point to each of the spheres. Ask: *What's the same about all of these spheres? What is different?* Reinforce the word 'sphere'.

- Repeat for cubes, cuboids and cylinders.
- Display the Shape set Digital Tool. Place a sphere onto the screen. Ask: *What is this shape?*
- Hold up a 3D geometric sphere, to help children recognise the picture of the sphere on screen. Say: *Here is a sphere. This is a picture of a sphere. It's what it looks like in a book or on the screen.*
- Place a cube next to the sphere. Ask: *What is this shape called?*
- Hold up a 3D geometric cube, next to the picture on the screen. Say: *Here is a cube. This is a picture of a cube.*
- Point to the sphere and cube on the screen. Ask: *What is the same about them?* Discuss responses, e.g. they are about the same size, they are both 3D, solid shapes.
- Ask: *What is different?* Discuss responses, e.g. a sphere is round, a sphere has 1 face/surface, a sphere rolls; a cube has more faces than a sphere, a cube has 6 faces, a cube has all square faces, a cube has all flat faces, a cube slides.
- Click 'Clear all'. Repeat to compare other pairs of shapes: sphere and cylinder; sphere and cuboid; cylinder and cube; cylinder and cuboid; cube and cuboid.

Explore in groups

Collect the four shapes

- Before the activity, prepare some 'feely' bags. Place 12 3D geometric shapes in each bag: 2 cubes, 2 cuboids, 2 cylinders, 2 spheres and 4 other shapes. Invite children to take a bag and feel inside for a shape. Ask them to try to take out a cube, cuboid, cylinder and sphere, without removing any other shapes.

What's the same/different?

- On a table, place a cube, cuboid, cylinder and sphere card from PCMs 25 and 26, a cube, cuboid, cylinder and sphere 3D geometric shape, and a real-world example of a cube, cuboid, cylinder and sphere. Invite children to work in pairs. Each child takes a card, geometric shape or real-world object and says the shape. Ask children to make statements comparing the two shapes: *What is the same, and what is different about them?*

Teaching notes

Collect all the shapes

- Shuffle the 6 sphere cards from PCM 26 and 6 cube cards from PCM 25. Spread them out face down. Encourage children to play a game in pairs. One child will collect spheres, the other will collect cubes. They take turns to turn over a card. If the card shows the shape they are collecting, they keep the card. If not, they put it back. The winner is the first child to collect their 6 cards.

Variations: use a different pair of shapes using other cards from PCMs 25 and 26, e.g. spheres and cylinders; cylinders and cubes; cubes and cuboids.

Session 7: Sort 3D shapes

Children sort 3D shapes into groups.

You will need:

3D geometric shapes; real-world examples of cubes, cuboids, cylinders and spheres; PCMs 25 and 26; tray; piece of fabric; four sorting hoops; labels: 'cubes', 'cuboids', 'cylinders' and 'spheres'; non-transparent ('feely') bags; large books; sand tray; 4 dishes; coloured pencils or crayons; *Maths Foundation Activity Book C*

Getting started

- Before the session, place several 3D geometric shapes and real-world examples of cubes, cuboids, cylinders and spheres on a tray. Cover it with a piece of fabric. Keep one of each 3D geometric shape to use at the start of *Teaching*.
- Begin by holding up shape cards from PCMs 25 and 26. Ask children to name each shape.

Teaching

- Hold up a 3D geometric sphere. Ask: *What is this shape called? How do you know it's a sphere?* Discuss the properties: a 3D, solid shape that is round, with 1 face/surface like a ball. It rolls.
- Place a sorting hoop on the floor. Label it: 'spheres'. Place the sphere in the hoop.
- Repeat for cube, cylinder and cuboid.
- Show children the covered tray containing cubes, cuboids, cylinders and spheres.

- Choose a child. Say: *Without looking, feel for a shape under the fabric and tell me its name.* The child does so, then holds up the shape. Ask: *Is Davian right?* Ask the child to place the shape into the correct hoop.
- Repeat with different children.
- If appropriate, repeat, sorting the shapes into three groups: 'slide', 'roll' and 'slide and roll'.
- Count how many of each shape there are. Briefly discuss the similarities and differences between the shapes.

Explore in groups

Name the shape

- Before the activity, prepare some 'feely' bags. Put a selection of 3D geometric shapes in each bag (cubes, cuboids, cylinders, spheres). Provide large books that can stand upright to create a screen. Encourage children to play a game in groups. They take turns to secretly take a shape out of their group's bag. They place it behind a book and slowly reveal the shape. The other children have to name the shape as soon as they can.

Sort the shapes

- Provide a cube, cuboid, cylinder and sphere 3D geometric shape, and a real-world example of a cube, cuboid, cylinder and sphere. Encourage children to sort the shapes. How many different ways can they sort the shapes into two, three or four groups?

Sand station

- Before the activity, bury some 3D geometric shapes in the sand. Include spheres, cylinders, cubes, cuboids and a few other shapes. Provide four dishes. Let children hunt for the shapes. When they have found some, encourage them to start sorting them into groups using the dishes. Ask them to tell you about the shapes they have found and how they have chosen to sort them.

Maths Foundation Activity Book C (adult-led)

Page 22 – Sort

Session 8: Make models using 3D shapes

Children use 3D shapes to make models.

You will need:

PCMs 25 and 26; construction materials, e.g. building blocks and recycled boxes and packaging, especially cubes, cuboids, cylinders and spheres; glue or tape; paint, water, paintbrushes, aprons; paper; coloured pencils or crayons; *Maths Foundation Activity Book C*; pencils

Getting started

- Before the session, set out all the building blocks and recycled boxes and packaging.
- Begin by holding up shape cards from PCMs 25 and 26. Ask children to name each shape.

Teaching

- Explain: *Today we're going to be making models using different shapes. You could make buildings, or bridges, or anything you like.*
- Start to build a model. Take one building block/recycled box/packaging at a time from the collection. As you take each resource, discuss with the children where might be a good place to put it. Emphasise which shapes slide (cube, cuboid, cylinder), roll (sphere, cylinder) and stack (cube, cuboid, cylinder). Ask: *Do you think this sphere will stay on top of this cuboid? What will happen to it? Why?*
- Continue until you have built a structure using five or six resources. Discuss the model with the children, e.g. *What shape is next to the cuboid? What shape is on top? What shape is below it? What other shape could I have put here? Might that be better? Why?*
- Start a new structure with the children's help. At each stage, ask them to tell you which shape you should use next and why, and where you should place it. Encourage children to use accurate shape and position vocabulary.

Explore in groups

Construction station

- Provide lots of building blocks and/or recycled boxes and packaging (cubes, cuboids, cylinders, spheres). Let children build models

using the resources. If appropriate, allow them to glue or tape the resources together and paint/decorate their model. Ask children to tell you about their model.

Construction station

- Shuffle the 8 geometric shape cards from PCMs 25 and 26. Place them face down in a pile. Provide a selection of different building blocks and recycled boxes and packaging (cubes, cuboids, cylinders, spheres). Encourage children to build a model, using the cards to guide them. Ask them to turn over a card and take a matching resource. They keep doing this in order to build their model. If appropriate, allow children to glue or tape the resources together and paint/decorate their model. Ask children to tell you about their model.

Painting and drawing station

- Encourage children to draw a picture of any model(s) they have made (or they could draw one made by someone else).

Maths Foundation Activity Book C (adult-led)

Page 23 – Slide or roll

Assessment opportunities

Assess children's learning against the objectives for this unit, using the guidance on formative assessment on pages 24–25, and record your observations in the Unit 15 progress tracking grid on page 33. The relevant pages of *Activity Book C* can also be used for assessment.

Can the children:

- recognise and name cubes, cuboids, cylinders and spheres in different sizes and from different perspectives, as well as real-world examples?
- compare two 3D shapes?
- sort 3D shapes into groups?
- use 3D shapes to make models?

Objects cards 1–10

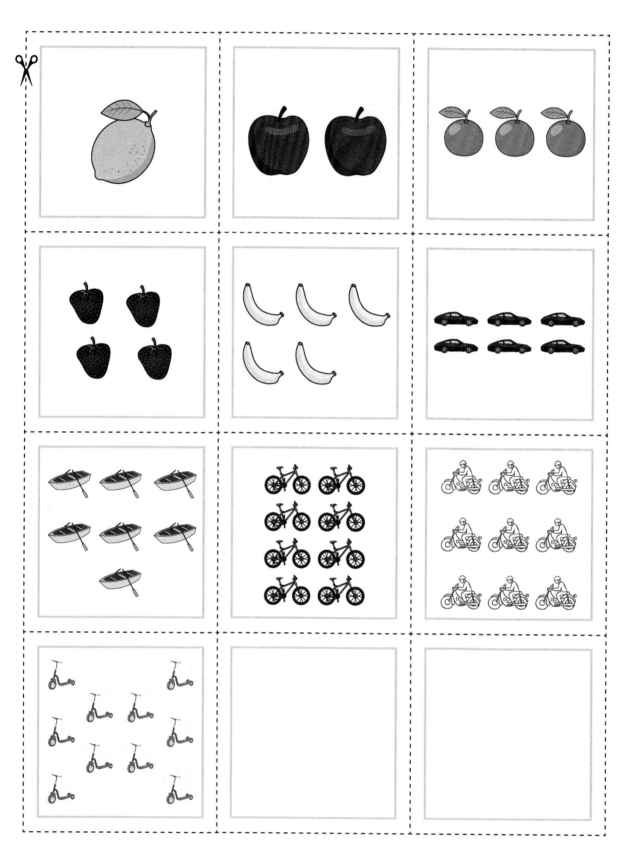

Fingers cards 1–10

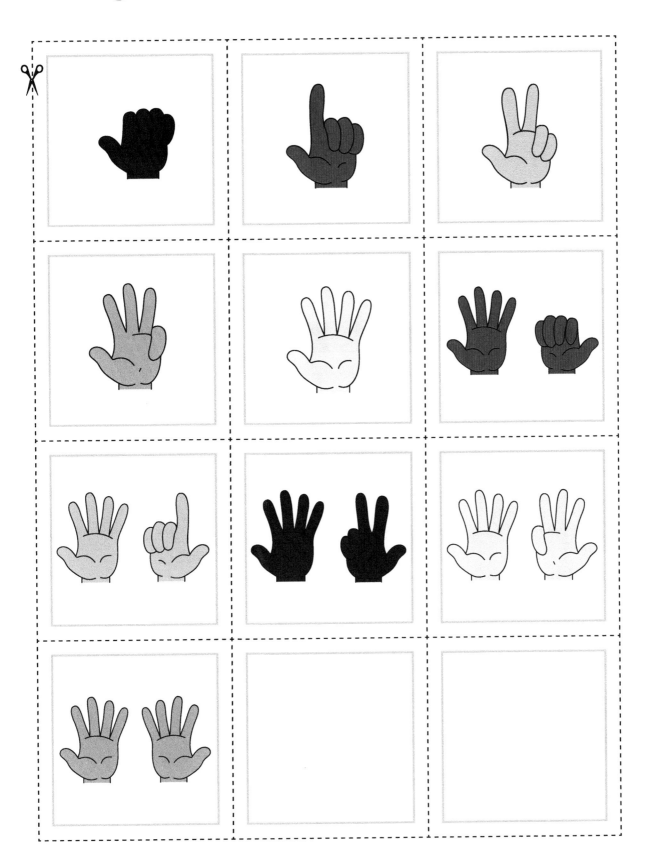

Cubes cards 1–10

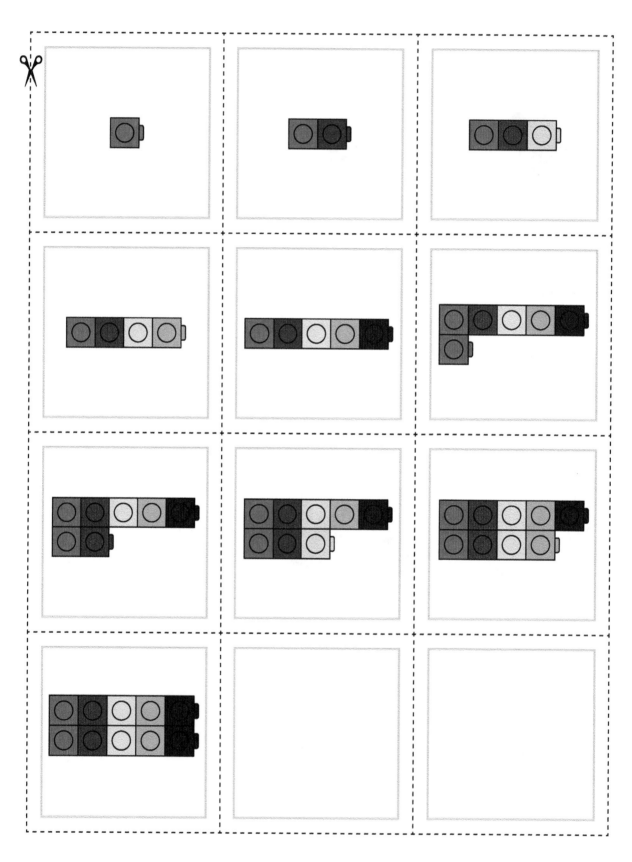

Ten-frame cards 1–10

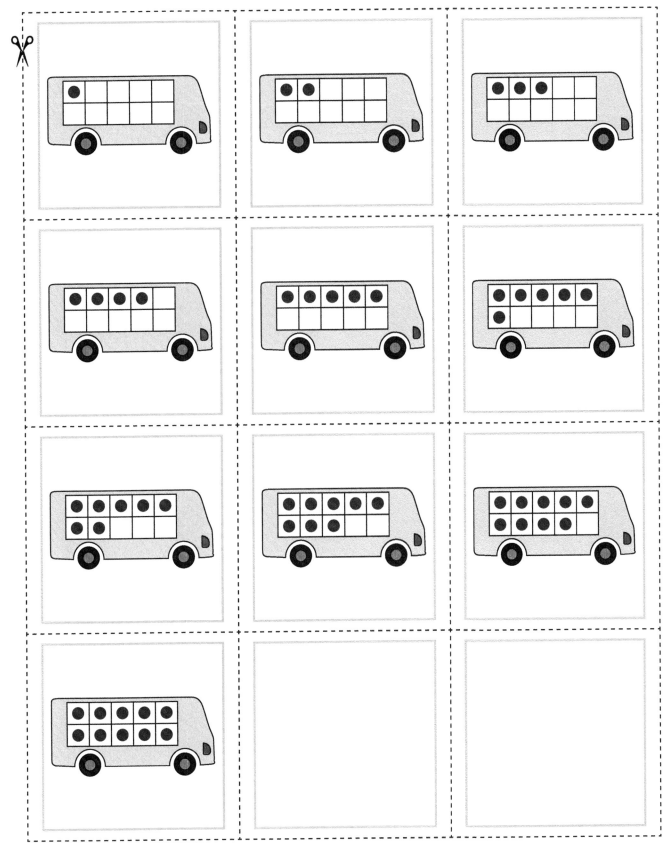

Numeral cards 1–10

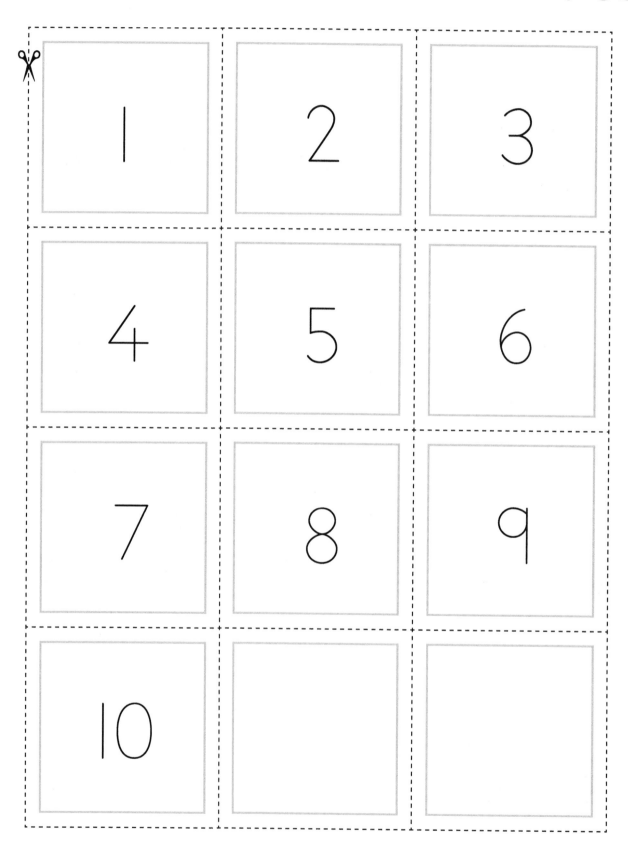

Name: _____   Date: _____

Two trees

Name: _____   Date: _____

Trace and write 1 to 5

Name: _____   Date: _____

Trace and colour 1 to 5

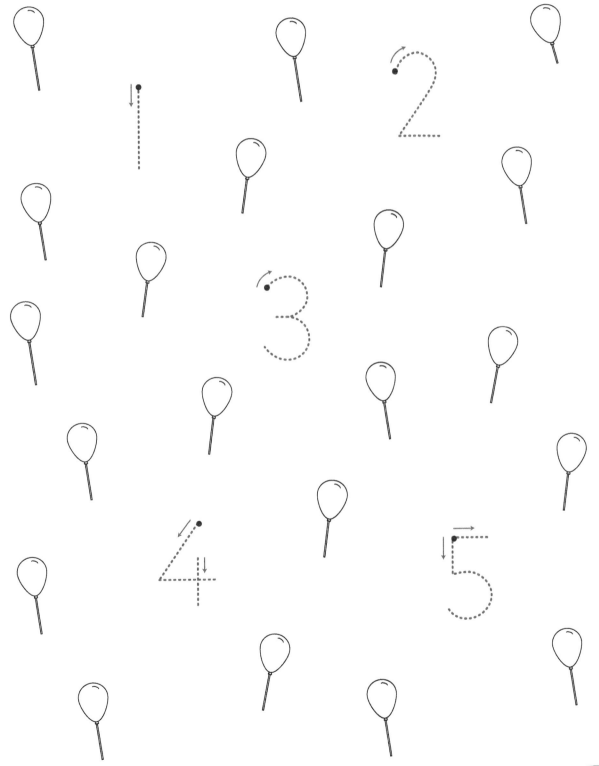

Subitising cards 1–6 set A (dots in regular patterns)

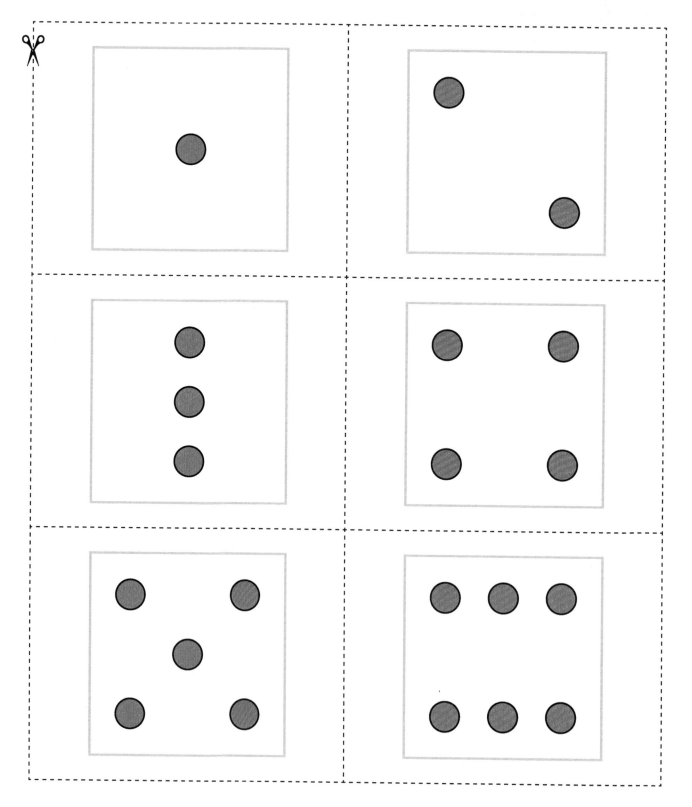

Subitising cards 1–6

set B (dots in irregular formations)

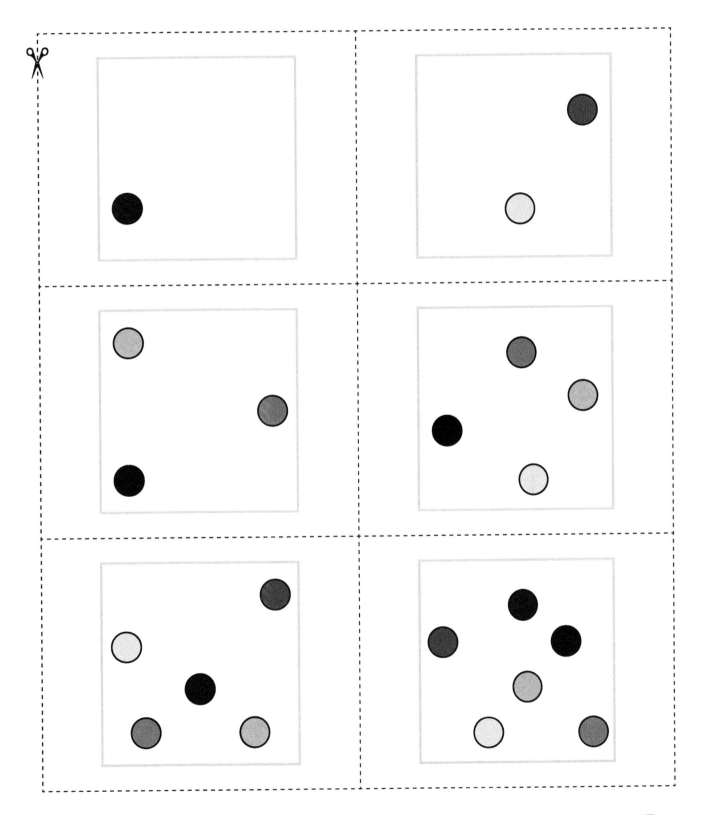

Blank ten-frame cards

Name: _____   Date: _____

Trace and write 6 to 10

Name: _____   Date: _____

Trace and colour 6 to 10

Ladybirds

1–10 number track

1	2	3	4	5	6	7	8	9	10
1	2	3	4	5	6	7	8	9	10
1	2	3	4	5	6	7	8	9	10

Counters cards 5–10

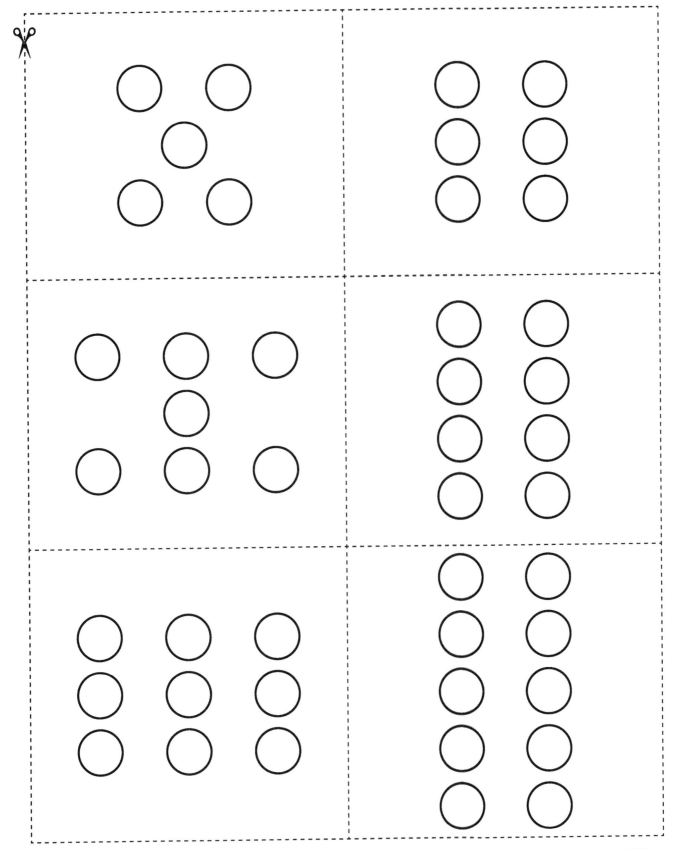

Direction track

Instructions:

To create a Direction track, cut along the dashed lines only, and glue the boxes indicated underneath the last box in the previous column.

		glue
	glue	

Circles

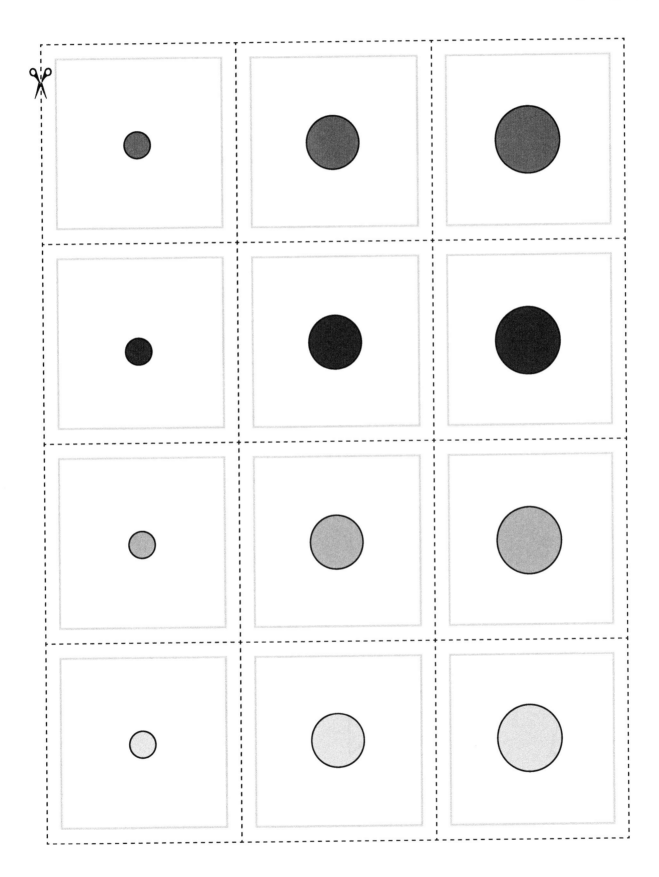

Triangles

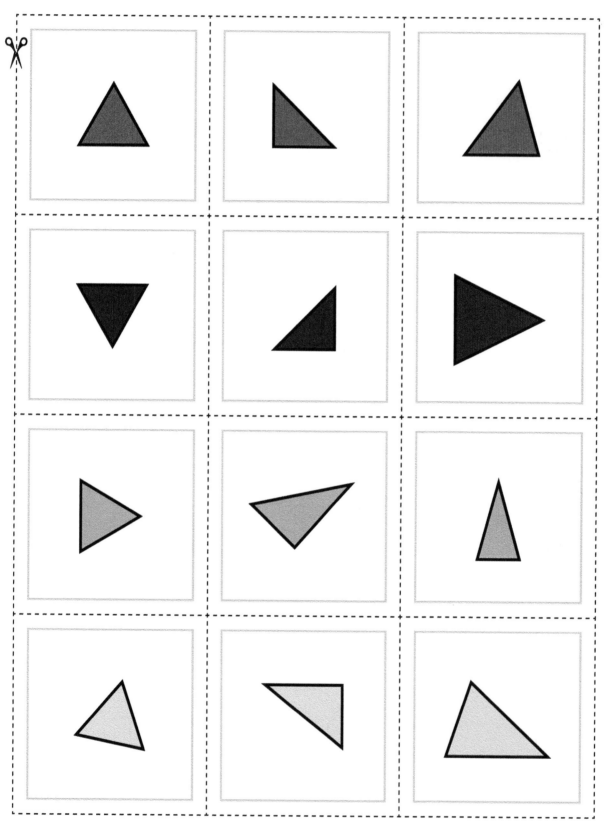

Squares

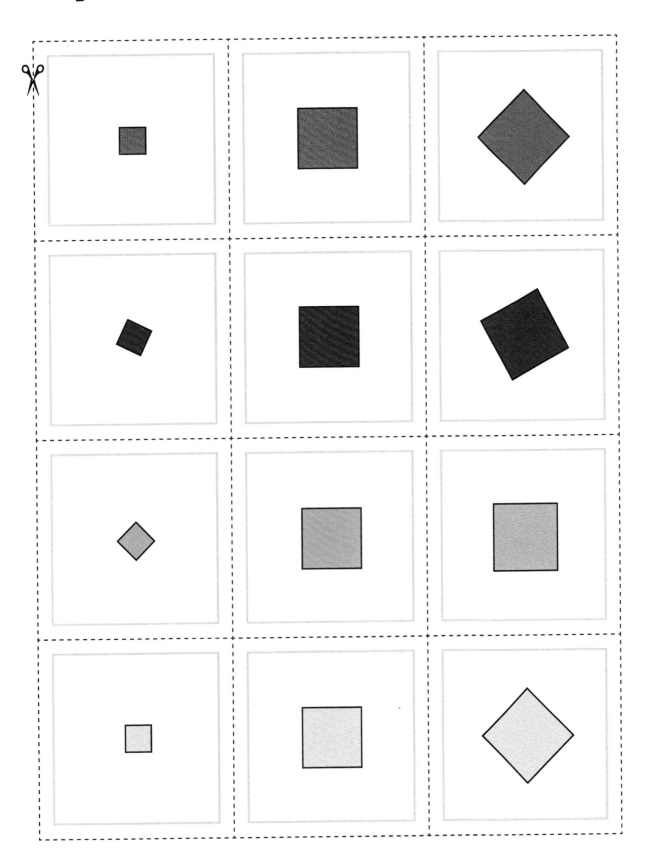

185

Rectangles

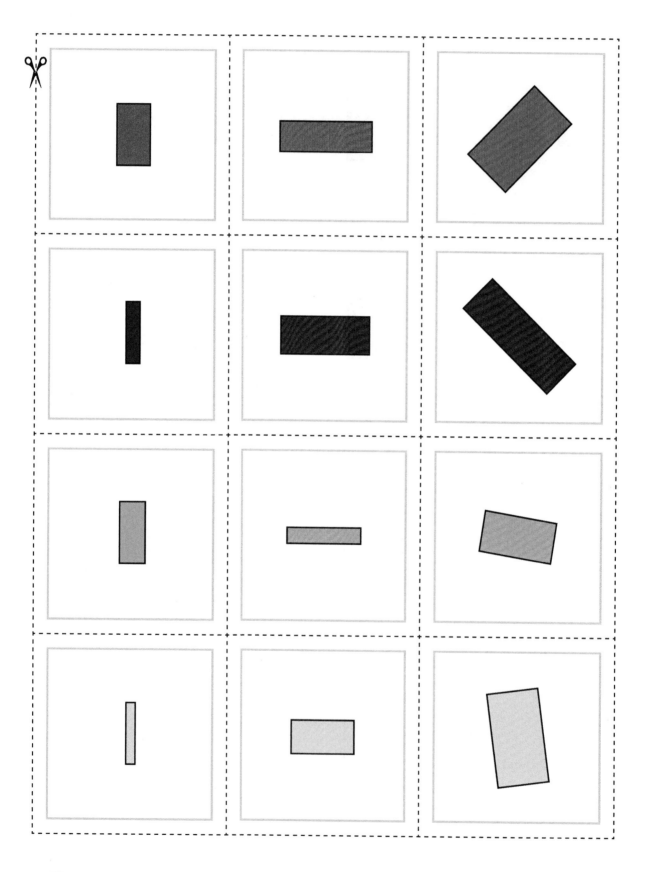

Name: _____   Date: _____

2D shapes

Pattern track (A)

Pattern track (B)

Cubes and cuboids

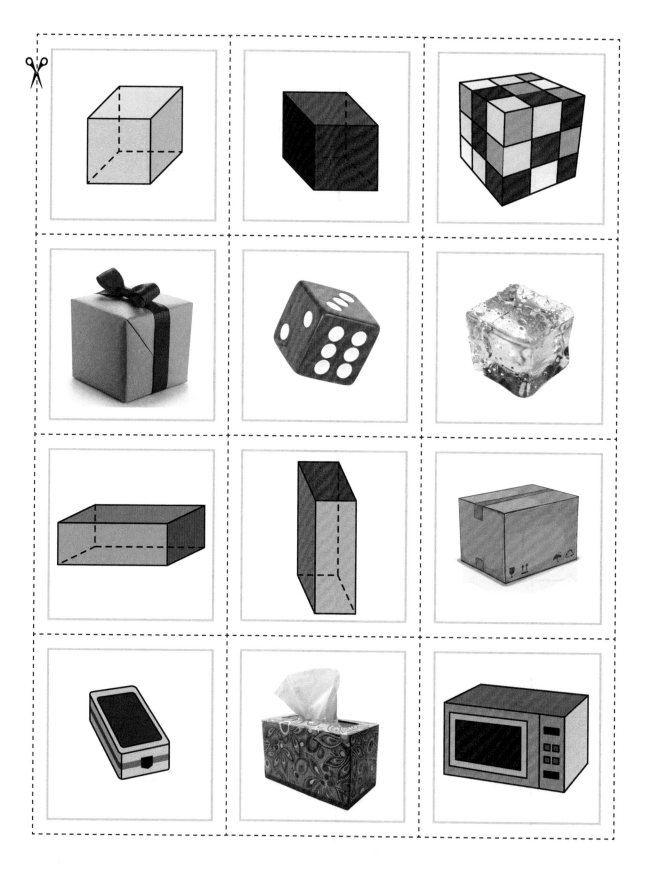

Cylinders and spheres

Glasses and bowls

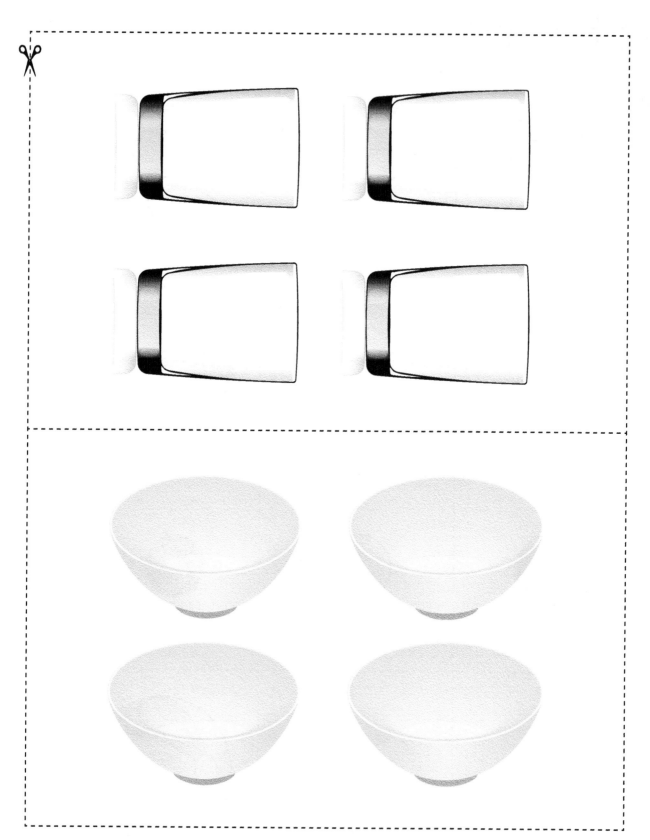

Asha's day

The Enormous Turnip

Days of the week

Monday	Friday
Tuesday	Saturday
Wednesday	Sunday
Thursday	

Luke's exercise diary

Hands-on (1 to 10)

8

4

1

9

3

2

5

10

7

6

Number fluency games and activities

The Number fluency games and activities are *not* intended to teach children new concepts. They are designed for children to practise and consolidate concepts previously taught. It is important to be aware of the number range that children have been taught so far, and not to extend the activity to include numbers that go beyond this.

- Units 1 to 5 focus on numbers 1 to 5.
- Units 6 to 15 focus on numbers 1 to 10.

For this reason, you will need to modify some of the activities slightly in order to keep within the appropriate number range.

The same also applies for the types of questions asked when using the addition and subtraction activities.

It is strongly advised that classrooms display number bunting or a large number track showing numerals and a matching picture for each numeral (e.g. 4 dots below the number 4). It is recommended that the number bunting/track shows numbers 1–5 when teaching Units 1 to 5, and numbers 1–10 for Units 6 to 15.

Counting and understanding numbers

1 Finger counting

- Count on in ones together from 1 to 10, starting with closed fists and putting up 1 finger (starting with the thumb) as each number is said, so that all 10 fingers are held up on the count of 10.
- Then count back in ones from 10 to 1, folding down 1 finger as each number is said.
- Repeat several times, speeding up as appropriate.

2 Show me

- Hold up two fingers. Say: *Two*. Hold up four fingers. Say: *Four*. Repeat for other numbers from 1 to 10.
- Then say: *I'm going to say a number. I want you to quickly hold up that number of fingers.*
- In any order, say each of the numbers 1 to 10 in turn.
- Repeat several times.

3 Continue counting

- Slowly start counting on from 1 to 10: *1, 2, 3, …*
- Point to a child who slowly continues the count.
- Then point to another child. The first child stops counting and the second child continues the count.
- When 10 is reached the count goes back to 1.

Variation: when 10 is reached, children count backwards.

4 Animal counting

- Children sit on the floor in a circle.
- They count round the circle, starting from 1 by choosing an animal and making the appropriate number of animal noises. For example, the first child barks once, the second child miaows twice, the third child chirps 3 times, and so on up to 10.
- The eleventh child starts again from 1.

Variations:

- After the tenth child, the next child starts counting backwards from 10 (so the eleventh child makes 9 animal noises).
- Instead of making animal noises, children make other noises, e.g. clap their hands, or pat their knees.
- Instead of making noises, children perform simple actions, e.g. hands on head, or jump in the air.

5 Next or before

- Children sit on the floor in a circle.
- Choose a child, and say a number less than 10, e.g. 7.
- That child then says the next number, e.g. 8.
- That child then chooses another child and says a different number less than 10, e.g. 3.
- The game continues with children choosing a child and saying a number less than 10.

Variation: children say the number before.

6 Say the target number

- Children sit on the floor in a circle.
- Choose a target number, e.g. 6. (Perhaps write it on the board or hold up a number card to help children remember the number.)
- Count on around the circle from 1, with children taking turns to say the next number in the count.
- The child who says the target number stands up.
- When 10 is reached, start counting again from 1.
- Continue until there are five children standing.

Variation: count backwards.

7 Fizz

- Children sit on the floor in a circle.
- Choose a number, e.g. 4. (Perhaps write it on the board or hold up a number card to help children remember the number.)
- Count on around the circle from 1, with children taking turns to say the next number in the count.
- When the chosen number is reached, instead of saying the number, the child whose turn it is says 'Fizz'.
- Continue counting around the class for as long as is appropriate.

Variation: count backwards.

8 Say it and you're out

- Children stand in a circle.
- Write three numbers on the board, e.g. 2, 5, 9.
- Count on around the circle from 1, with children taking turns to say the next number in the count.
- When a child says one of the numbers written on the board, the child sits down and is out.
- Continue counting around the class until one child is left. That child is the winner.

Variation: count backwards.

9 Counting stick

You will need: a 1–10 counting stick (a stick ideally 1 metre in length, marked in 10 equal divisions, with the divisions labelled 1–10, e.g. using sticky notes)

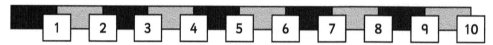

- Pointing to the numbers on the counting stick, count on from 1 to 10.
- As children become more confident:

  - count backwards from 10 to 1
  - remove some of the numbers: *Which number belongs here?*
  - gradually remove almost all of the numbers leaving only 1 and 10.

10 Who's left standing? (counting objects)

You will need: PCM 1 (objects cards 1–10)

- Children stand.
- Give each child a 1 to 10 objects card.
- Say: *Look at your card. Sit down if your card shows … things* (e.g. 3 things). Repeat for eight of the other numbers (e.g. all the numbers except 9), in a random order.
- The winners are all the children still standing with e.g. 9 things on their card.

Variations:

- Use PCM 2 (fingers cards) and say: *Sit down if your card shows … fingers.*
- Use PCM 3 (cubes cards) and say: *Sit down if your card shows … cubes.*
- Use PCM 4 (ten-frame cards) and say: *Sit down if your card shows … dots.*

11 Target board (counting objects)

You will need: Target board slide

- Display the slide.
- Ask questions similar to the following:

  - *How many stars are there?*
  - *Which object are there 6 of?*
  - *Are there more flowers or crayons?*
  - *Are there fewer hats or turtles?*

12 Marbles in a tin

You will need: marbles; tin (or similar)

- Show children the empty tin.
- Tell children to close their eyes.
- Slowly, and one at a time, drop from 1 to 10 marbles into the tin.
- Tell children to open their eyes and tell you how many marbles are in the tin.
- Slowly empty the marbles from the tin one at a time and count how many there are.
- Repeat several times.

Variation: use other suitable readily available resources.

Reading and writing numbers

13 Circle count

You will need: PCM 5 (numeral cards 1–10)

- Children sit in a circle on the floor.
- Shuffle the number cards. Place them face down in a pile in the centre of the circle.
- Choose a child to stand up and take the top card from the pile.
- That child then counts round the circle that number of children, starting from the child who was sitting next to them.
- They then change places with the child they are pointing to when they say the last number in the count. They give the number card to you.
- The 'new' child takes the next card from the pile and counts round from where they were sitting.
- Repeat until all the number cards have been used.
- If appropriate, and time allows, reshuffle the cards and continue.

14 Match the cards

You will need: PCM 1 (objects cards 1–10); PCM 5 (numeral cards 1–10)

- Shuffle the numeral cards and place them face down in a pile.
- Give each child an objects card.
- Turn over the top numeral card and show the number to the children.
- All the children with the card that shows the matching number of objects hold up their card.
- Repeat until all the numeral cards have been revealed and matched.

Variations:

- – Give each child a fingers card from PCM 2.
- – Give each child a cubes card from PCM 3.
- – Give each child a ten-frame card from PCM 4.
- – Use the Number cards Digital Tool instead of the numeral cards from PCM 5.
- – Alter the activity by giving each child a numeral card from PCM 5 and displaying the corresponding Dot cards or Ten-frame cards from the Number cards Digital Tool.

15 Who's left standing? (counting on and back/reading numbers)

You will need: PCM 5 (numeral cards 1–10)

- Children stand.
- Give each child a numeral card.
- Say: *Look at your card. Sit down if the number on your card is the next number in the pattern:*

  - 2, 3, 4, (5)   - 5, 6, 7, (8)   - 4, 3, 2, (1)
  - 7, 6, 5, (4)   - 7, 8, 9, (10)   - 6, 5, 4, (3)
  - 6, 7, 8, (9)   - 5, 4, 3, (2)   - 4, 5, 6, (7).

- The winners are all the children still standing with 6 on their cards.

Variation: choose a different number to be 'left standing' and adapt the questions so that your chosen number is never the answer.

Comparing and ordering numbers

16 Point to it (numbers)

You will need: Point to it slides; two rulers (or something similar to point with)

- Before the activity, decide which of the three slides to display. All three slides show numbers 1–10, but in slightly different formats and levels of challenge.
- Ask two children to stand either side of the numbers. Give each of them a ruler.
- Give instructions similar to the following:

  - *Point to the number 5.*
  - *Point to the next number in this pattern: 6, 7, 8, 9, …*
  - *Point to the next number in this pattern: 5, 4, 3, 2, …*
  - *Point to any number less than 4.*
  - *Point to any number more than 6.*
  - *Point to the number that comes before 6.*
  - *Point to the number that comes after 7.*

- The first child to point to the correct number stays in. The other child sits down.
- Choose another child to take the place of the child that sat down.
- Continue as above.
- Which child can stay in the longest?

17 Hands-on (numbers)

You will need: PCM 32 (Hands-on (1 to 10))

- Arrange children into pairs sitting at tables. Give each pair a copy of PCM 32.
- Tell children to put the sheet between them, and each put their index finger on a hand at the bottom of the game board.
- Give instructions similar to the following:

  - *Point to the number 3.*
  - *Point to the next number in this pattern: 4, 5, 6, 7, …*

> – *Point to the next number in this pattern: 10, 9, 8, 7, ...*
> – *Point to the number that comes before 3.*
> – *Point to the number that comes after 5.*

- Children race to be the first one in their pair to point to the correct number.
- Ask a child: *What number did you point to?* Ask the other children: *Is that the right number? Who else pointed to that number first?*
- If able, children can keep count of how many rounds they win.
- Play for as long as is appropriate.
- The winner in each pair is the child who wins more rounds.

18 Change places

You will need: PCM 5 (numeral cards 1–10)

- Children sit in a circle on the floor.
- Give each child a number card. Say: *Look at the number on your card. If your number is more than 7, change places with another child.*
- Children with cards showing 8, 9 or 10 stand and swap places.
- Repeat with other more than/less than instructions.

19 Remember the numbers

You will need: PCM 5 (numeral cards 1–10); easel

- Display three number cards on the easel, e.g. 2, 5 and 8.
- Tell children to look carefully at the numbers and try to remember them.
- After a suitable length of time, turn over each of the number cards.
- Ask questions similar to the following:

> – *Can you remember one of the numbers?*
> – *Can you tell me another number?*
> – *What number is less than 4?*
> – *What number is more than 6?*

Variation: use the Number cards Digital Tool instead of the numeral cards from PCM 5. When placing the numeral cards onto the work area ensure that the numbers are not in order and 'Hide all values'.

20 Guess my number

You will need: PCM 5 (numeral cards 1–10), or sticky notes or small pieces of paper; tape; pen

- Choose a number card (or write a number on a sticky note), e.g. 5.
- Stick the number on a child's back without them seeing the number. Make sure that the rest of the children see the number.
- Invite children to give clues about the number, but without saying the number name, e.g. *It is more than 2. It is less than 8. It has straight lines and a curved line. It's one of the numbers on that board.*
- Once the child has guessed the number, repeat with another child.

21 More or less?

You will need: PCM 5 (numeral cards 1–10); easel/whiteboard or washing line and 10 pegs

- Shuffle the number cards. Either spread them out face down along the length of the easel/whiteboard, or hang the cards on the washing line. Do not let children see the numbers.
- Tell children that each card has a number on it from 1 to 10.
- Turn over the first card (the one on the left, as viewed by the children) to reveal the number.
- Ask children to decide whether they think the number on the next card will be more or less than the current number.
- If they think the number will be more, they put their hands in the air.
- If they think the number will be less, they put their hands on their head.
- Turn over the second card to reveal the number.
- Keep going until all the numbers are revealed.
- With the children's help, order the number cards, smallest to largest.
- End by reciting the numbers from smallest to largest, and, if appropriate, largest to smallest.

Understanding addition and subtraction

22 Adding fingers

- Choose two children to come and stand with you in front of the class.
- These children each put one hand behind their backs.
- Say: *Get ready! Show!*
- On *Show!* the two children bring their hands from behind their backs with 1, 2, 3, 4 or 5 fingers standing up.
- The rest of the children count the total number of fingers standing up.
- Ask a child from the class for the total. Confirm the answer.
- Repeat until all children have had a turn in front of the class.

23 Adding cards

You will need: Number cards Digital Tool

- Display the Number cards Digital Tool. In Set up, select the Number range '1-5', the Type of cards 'Dot cards'. Hide the card values. Place the five cards onto the work area.
- Click on two of the cards to show the dots. Ask: *How many dots are on this card? How many dots are on this card? How many is this altogether?*
- Click on the two cards to hide the dots.
- Repeat several times.

Variation: for Type of cards, select 'Ten-frame cards' or 'Numeral cards' instead of 'Dot cards'.

24 Subtracting cards

You will need: PCM 1 (objects cards 1–10); easel

- Choose two cards and place them on the easel.
- Ask questions such as: *How many … are on this card? How many … are on this card? Are there more/fewer … or …? How many more/fewer are there?*

- Remove one of the cards from the easel. Point to the remaining card and ask: *How many ... are on this card? If I took away ... of these ..., how many ... would be left?*
- Repeat for other cards.

Variations:

– Use PCM 2 (fingers cards).
– Use PCM 3 (cubes cards).
– Use PCM 4 (ten-frame cards).
– Use the 'Dot cards' or 'Ten-frame cards' from the Number cards Digital Tool instead of the cards from the PCMs.

25 Target board (addition and subtraction)

You will need: Target board slide

- Display the slide.
- Ask questions similar to the following: *How many hats are there? How many cars are there? How many hats and cars are there altogether? How many stars are there? How many kites are there? Are there more stars or kites? How many more stars are there?*

26 Who's left standing? (1 more and less)

You will need: PCM 5 (numeral cards 1–10)

- Children stand.
- Give each child a numeral card.
- Say: *Look at your card. Sit down if the number on your card is:*

– *1 more than 2 (3)*	– *1 more than 5 (6)*	– *1 less than 10 (9)*
– *1 less than 3 (2)*	– *1 more than 4 (5)*	– *1 less than 9 (8)*
– *1 more than 6 (7)*	– *1 more than 9 (10)*	– *1 less than 2 (1).*

- The winners are all the children still standing with 4 on their cards.

Variations:

– Choose a different number to be 'left standing' and adapt the questions so your chosen number is never the answer.
– Use numeral cards 2–10 and only ask '1 more than' questions.
– Use numeral cards 1–9 and only ask '1 less than' questions.

27 Who's left standing? (addition and subtraction)

You will need: PCM 5 (numeral cards 1–10)

Game 1: Addition facts to 10

- Children stand.
- Give each child a 2 to 10 numeral card.
- Say: *Look at your card. Sit down if the number on your card is the answer to:*

– *2 plus 1 (3)*	– *6 plus 2 (8)*	– *8 add 1 (9)*
– *4 plus 2 (6)*	– *2 more than 3 (5)*	– *5 add 2 (7).*
– *1 more than 9 (10)*	– *2 plus 2 (4)*	

- The winners are all the children still standing with 2 on their cards.

Game 2: Subtraction facts to 10

- Children stand.
- Give each child a 1 to 9 numeral card.
- Say: *Look at your card. Sit down if the number on your card is the answer to:*

  - *9 subtract 2* (7)
  - *3 take away 1* (2)
  - *1 less than 9* (8)
  - *1 less than 10* (9)
  - *7 take away 1* (6)
  - *3 take away 2* (1)
  - *6 subtract 2* (4)
  - *5 subtract 2* (3).

- The winners are all the children still standing with 5 on their cards.

Game 3: Addition and subtraction facts to 10

- Children stand.
- Give each child a 1 to 10 numeral card.
- Say: *Look at your card. Sit down if the number on your card is the answer to:*

  - *5 plus 2* (7)
  - *2 subtract 1* (1)
  - *9 add 1* (10)
  - *2 more than 4* (6)
  - *8 plus 1* (9)
  - *5 subtract 2* (3)
  - *4 take away 2* (2)
  - *7 take away 2* (5)
  - *2 plus 2* (4)

- The winners are all the children still standing with 8 on their cards.

Variation: choose a different number to be 'left standing' and adapt the questions so your chosen number is never the answer.

28 Point to it (addition and subtraction)

You will need: Point to it slides; two rulers (or something similar to point with)

- Before the activity, decide which of the three slides to display. All three slides show numbers 1–10, but in slightly different formats and levels of challenge.
- Ask two children to stand either side of the numbers. Give each of them a ruler.
- Ask an appropriate addition or subtraction question where the answer is on the slide, e.g. *What is 1 more than 3?*
- The first child to point to the correct number stays in. The other child sits down.
- Choose another child to take the place of the child that sat down.
- Continue as above.
- Which child can stay in the longest?

29 Hands-on (addition and subtraction)

You will need: PCM 32 (Hands-on (1 to 10))

- Arrange children into pairs sitting at tables. Give each pair a copy of PCM 32.
- Tell children to put the sheet between them, and each put their index finger on a hand at the bottom of the game board.
- Ask an appropriate addition or subtraction question where the answer is on the game board, e.g. *What is 5 take away 1?*
- Children race to be the first one in their pair to point to the correct answer.

- Ask a child: *What's the answer?* Ask the other children: *Is that the right answer? Who else pointed to that answer first?*
- If able, children can keep count of how many rounds they win.
- Play for as long as is appropriate.
- The winner in each pair is the child who wins more rounds.

30 Around the class

- Children either sit in a circle or at their tables. There must be a logical path from one child/table to the next.
- Ask one child to go and stand behind the child who is sitting next to them.
- Ask an appropriate question to these two children, e.g. *What is 1 more than 3?*
- If the child who is standing calls out the correct answer first, they move and stand behind the next child.
- If the child that is sitting calls out the correct answer first, they stand up and go behind the next child. The child that was standing sits in the place of the child they were just standing behind.
- The game continues in this way around the room.
- When you have gone around the class say: *All change!* and children can go back to their original positions.

Patterns and sequences

31 Patterns

You will need: container of counting apparatus in different colours (e.g. counters, beads, interlocking cubes)

- Sit with the children on the floor in a circle.
- Make a simple AB pattern by:
  - taking a counting object from the container and placing it in front of you, e.g. red counter
  - placing a second object in front of the child next to you, e.g. a blue counter
  - placing a third object in front of the next child, e.g. a red counter
  - placing a fourth object in front of the next child, e.g. a blue counter.

- Pass the container to the next child and ask them to continue the pattern by choosing an object from the container and placing it in front of them, e.g. a red counter.
- Continue around the class until there is an object in front of each child.

Variations:

- Use other counting apparatus to make different AB patterns based on different criteria, such as shape, colour, size and type.
- Make ABC or ABB patterns.
- Make repeating patterns based on actions or sounds, e.g.
  - AB: clap, tap head, clap, tap head, …
  - ABC: hands on shoulders, hands on head, hands in the air, hands on shoulders, hands on head, hands in the air, …
  - ABB: clap once, tap head twice, tap head twice, clap once, tap head twice, tap head twice, …

Rhymes and songs

1 One finger, one thumb, keep moving

(While singing the song, move the body parts mentioned in the lyrics.)

One finger, one thumb, keep moving
One finger, one thumb, keep moving
One finger, one thumb, keep moving
We'll all be merry and bright.

One finger, one thumb, one arm, keep moving
One finger, one thumb, one arm, keep moving
One finger, one thumb, one arm, keep moving
We'll all be merry and bright.

One finger, one thumb, one arm, one leg, keep moving
One finger, one thumb, one arm, one leg, keep moving
One finger, one thumb, one arm, one leg, keep moving
We'll all be merry and bright.

One finger, one thumb, one arm, one leg, one nod of the head, keep
moving
One finger, one thumb, one arm, one leg, one nod of the head, keep
moving
One finger, one thumb, one arm, one leg, one nod of the head, keep
moving
We'll all be merry and bright.

We'll all be merry and bright.

2 One man went to mow

(Hold up the corresponding number of fingers as each number is said.)

One man went to mow, went to mow a meadow,
One man and his dog – woof! – went to mow a meadow.

Two men went to mow, went to mow a meadow,
Two men, one man and his dog – woof! – went to mow a meadow.

Three men went to mow, went to mow a meadow,
Three men, two men, one man and his dog – woof! – went to mow a
meadow.

Four men went to mow, went to mow a meadow,
Four men, three men, two men, one man and his dog – woof! – went to
mow a meadow.

Five men went to mow, went to mow a meadow,
Five men, four men, three men, two men, one man and his dog – woof! –
went to mow a meadow.

3 Here is the beehive

Here is the beehive

(Interlock fingers and clasp hands together to make a hive.)

But where are the bees?
Hidden inside where nobody sees.
Watch them come creeping out of the hive,
One, two, three, four, five.

(Look inside hive.)

(Release hands, move fingers on one hand around as if they are five bees flying.)

4 One elephant went out to play

(Hold up the corresponding number of fingers for each verse as the number is said.)

One elephant went out to play
Upon a spider's web one day.
He had such enormous fun
That he called for another elephant to come.

Two elephants went out to play
Upon a spider's web one day.
They had such enormous fun
That they called for another elephant to come.

(Repeat for three more verses: three, four, five elephants.)

All the elephants were out at play
Upon a spider's web one day.
They had such enormous fun
But there were no more elephants left to come!

5 This old man

(Hold up the corresponding number of fingers for each verse as the number is said.)

This old man, he played one,
He played knick-knack on my thumb.
With a knick-knack paddywhack,
Give a dog a bone,
This old man came rolling home.

This old man, he played two,
He played knick-knack on my shoe.
With a knick-knack paddywhack,
Give a dog a bone,
This old man came rolling home.

This old man, he played three,
He played knick-knack on my knee.
With a knick-knack paddywhack,
Give a dog a bone,
This old man came rolling home.

This old man, he played four,
He played knick-knack on my door.
With a knick-knack paddywhack,
Give a dog a bone,
This old man came rolling home.

This old man, he played five,
He played knick-knack on my hive.
With a knick-knack paddywhack,
Give a dog a bone,
This old man came rolling home.

This old man, he played six,
He played knick-knack on my sticks.
With a knick-knack paddywhack,
Give a dog a bone,
This old man came rolling home.

This old man, he played seven,
He played knick-knack up in heaven.
With a knick-knack paddywhack,
Give a dog a bone,
This old man came rolling home.

This old man, he played eight,
He played knick-knack on my gate.
With a knick-knack paddywhack,
Give a dog a bone,
This old man came rolling home.

This old man, he played nine,
He played knick-knack on my spine.
With a knick-knack paddywhack,
Give a dog a bone,
This old man came rolling home.

This old man, he played ten,
He played knick-knack once again.
With a knick-knack paddywhack,
Give a dog a bone,
This old man came rolling home.

6 There were two birds sitting on a stone

There were two birds sitting on a stone,
Fa, la, la, la, lal, de.
One flew away, and then there was one,
Fa, la, la, la, lal, de.
The other flew after, and then there was none,
Fa, la, la, la, lal, de.
And so the poor stone was left all alone,
Fa, la, la, la, lal, de.

7 Alice the camel

Alice the camel has five humps.
Alice the camel has five humps.
Alice the camel has five humps.
So go, Alice, go!
Boom, boom, boom, boom!

Alice the camel has four humps.
Alice the camel has four humps.
Alice the camel has four humps.
So go, Alice, go!
Boom, boom, boom, boom!

(Repeat for three more verses: three humps, two humps, one hump.)

Alice the camel has no humps.
Alice the camel has no humps.
Alice the camel has no humps.
Because Alice is a horse, of course!

8 Zoom, zoom, we're going to the moon

Zoom, zoom, zoom	*(Rub hands together while rocking back and forth.)*
We're going to the moon.	
Zoom, zoom, zoom	*(Rub hands together while rocking back and forth.)*
We're going to the moon.	
If you want to take a trip,	*('Walking' motion with two fingers.)*
Climb aboard my rocket ship.	
Zoom, zoom, zoom	*(Rub hands together while rocking back and forth.)*
We're going to the moon.	
5, 4, 3, 2, 1,	*(Hold 5 fingers up and count down.)*
Blast off!	*(Jump into the air.)*

9 Five little ducks

Five little ducks went swimming one day,	*(Hold up 5 fingers.)*
Over the hill and far away.	*(Wave.)*
Mama duck said: 'Quack, quack, quack, quack!'	*(Form hands like a beak, opening and closing fingers.)*
And only four little ducks came back.	*(Hold up 4 fingers.)*
Four little ducks went swimming one day,	*(Hold up 4 fingers.)*
Over the hill and far away.	*(Wave.)*
Mama duck said: 'Quack, quack, quack, quack!'	*(Form hands like a beak, opening and closing fingers.)*
And only three little ducks came back.	*(Hold up 3 fingers.)*
Three little ducks went swimming one day,	*(Hold up 3 fingers.)*
Over the hill and far away.	*(Wave.)*
Mama duck said: 'Quack, quack, quack, quack!'	*(Form hands like a beak, opening and closing fingers.)*
And only two little ducks came back.	*(Hold up 2 fingers.)*
Two little ducks went swimming one day,	*(Hold up 2 fingers.)*
Over the hill and far away.	*(Wave.)*
Mama duck said: 'Quack, quack, quack, quack!'	*(Form hands like a beak, opening and closing fingers.)*
And only one little duck came back.	*(Hold up 1 finger.)*
One little duck went swimming one day,	*(Hold up 1 finger.)*
Over the hill and far away.	*(Wave.)*
Mama duck said: 'Quack, quack, quack, quack!'	*(Form hands like a beak, opening and closing fingers.)*
And all her five little ducks came back.	*(Hold up 5 fingers.)*

10 Five little monkeys

Five little monkeys jumping on the bed,	*(Hold up 5 fingers and move them up and down.)*
One fell down and bumped his head,	*(Tap head.)*
Mama called the doctor and the doctor said:	*(Hand gesture mimicking a telephone call.)*
'No more monkeys jumping on the bed!'	*(Wagging the index finger.)*
Four little monkeys jumping on the bed,	*(Hold up 4 fingers and move them up and down.)*
One fell down and bumped his head,	*(Tap head.)*
Mama called the doctor and the doctor said:	*(Hand gesture mimicking a telephone call.)*
'No more monkeys jumping on the bed!'	*(Wagging the index finger.)*

(Repeat for two more verses: three monkeys, two monkeys.)

One little monkey jumping on the bed,	*(Hold up 1 finger and move them up and down.)*
She fell down and bumped her head,	*(Tap head.)*
Mama called the doctor and the doctor said:	*(Hand gesture mimicking a telephone call.)*
'Put those monkeys back to bed!'	*(Wagging the index finger.)*

11 Five currant buns

(Start with 5 fingers and after each currant bun is taken away, put down 1 finger.)

Five currant buns in a baker's shop,
Round and fat with a cherry on top.
Along came a boy with a penny one day,
Bought a currant bun and took it away.

Four currant buns in a baker's shop,
Round and fat with a cherry on top.
Along came a girl with a penny one day,
Bought a currant bun and took it away.

(Repeat for two more verses: three currant buns, two currant buns.)

One currant bun in a baker's shop,
Round and fat with a cherry on top.
Along came a boy with a penny one day,
Bought a currant bun and took it away.
Bought a currant bun and took it away!

12 Five little speckled frogs

(Start with 5 fingers and after each frog jumps into the pool, put down 1 finger.)

Five little speckled frogs,
Sat on a speckled log,
Eating some most delicious bugs – yum, yum!
One jumped into the pool,
Where it was nice and cool,
Then there were four speckled frogs – glug, glug!

Four little speckled frogs,
Sat on a speckled log,
Eating some most delicious bugs – yum, yum!
One jumped into the pool,
Where it was nice and cool,
Then there were three speckled frogs – glug, glug!

(Repeat for two more verses: three frogs, two frogs.)

One little speckled frog,
Sat on a speckled log,
Eating some most delicious bugs – yum, yum!
He jumped into the pool,
Where it was nice and cool,
Then there were no speckled frogs.

13 Mary at the cottage door

1, 2, 3, 4,
Mary at the cottage door,
Eating cherries off a plate,
5, 6, 7, 8.

14 One potato, two potatoes

(Hold your fists in front of you and stack them on top of each other starting with 'one potato', then with 'two potatoes' move the lower fist on top of the upper fist. Continue until 'ten potatoes'. At the end, when 'all' is sung, open both hands and shout 'all'.)

One potato, two potatoes, three potatoes – four.
Five potatoes, six potatoes, seven potatoes – more.
Eight potatoes, nine potatoes, ten potatoes – all.

One, two, three, four, five, six, seven, eight, nine, ten.

One potato, two potatoes, three potatoes – four.
Five potatoes, six potatoes, seven potatoes – more.
Eight potatoes, nine potatoes, ten potatoes – all.

15 One, two, three, four, five

One, two, three, four, five   *(Count on fingers.)*
Once I caught a fish alive.
Six, seven, eight, nine, ten   *(Count on fingers.)*
Then I let it go again.

Why did you let it go?   *(Hold hands with palms facing up.)*
Because it bit my finger so.
Which finger did it bite?
This little finger on the right.   *(Hold up little finger of right hand.)*

16 Ten little fishies

(Start with 1 finger and after each little fishy comes along, hold up another finger.)

One little fishy swimming in the sea
Along came another one, and then there were two.

Two little fishies swimming in the sea
Along came another one, and then there were three.

Three little fishies swimming in the sea
Along came another one, and then there were four.

Swimming, swimming, swimming in the sea
That big old shark will never catch me.

(Repeat for four, five and six fishies.)

Swimming, swimming, swimming in the sea
That big old shark will never catch me.

(Repeat for seven, eight and nine fishies.)

Swimming, swimming, swimming in the sea
That big old shark will never catch me.

Swimming, swimming, swimming in the sea
That big old shark will never catch me.

Swimming, swimming, swimming in the sea
That big old shark will never catch me.

He'll never
Never
Never
Never
That big old shark will never catch me!

17 Ten in the bed

(Start with 10 fingers and after each one falls out, put down 1 finger.)

There were ten in the bed,
And the little one said:
'Roll over! Roll over!'
So they all rolled over and one fell out.

There were nine in the bed,
And the little one said:
'Roll over! Roll over!'
So they all rolled over and one fell out.

There were eight in the bed,
And the little one said:
'Roll over! Roll over!'
So they all rolled over and one fell out.

(Repeat for six more verses: seven, six, five, four, three, two in the bed.)

There was one in the bed,
And the little one said:
'Alone at last!
Good night!'

18 Ten green bottles

(Start with 10 fingers and after each bottle falls, put down 1 finger.)

Ten green bottles sitting on the wall,
Ten green bottles sitting on the wall,
And if one green bottle should accidentally fall,
There'll be nine green bottles sitting on the wall.

Nine green bottles sitting on the wall,
Nine green bottles sitting on the wall,
And if one green bottle should accidentally fall,
There'll be eight green bottles sitting on the wall.

Eight green bottles sitting on the wall,
Eight green bottles sitting on the wall,
And if one green bottle should accidentally fall,
There'll be seven green bottles sitting on the wall.

(Repeat for six more verses: seven, six, five, four, three, two green bottles.)

One green bottle sitting on the wall,
One green bottle sitting on the wall,
And if one green bottle should accidentally fall,
There'll be no green bottles sitting on the wall.

19 A-counting we will go

(Sung to the tune of 'A-hunting we will go'.)

(Chorus)   A-counting we will go,
A-counting we will go,
Let's say each number one by one from 1 up to 10

Let's count to number 5,
Let's count to number 5.
Let's count the ducklings up to 5.
1, 2, 3, 4, 5.

(Repeat chorus.)

Let's count to number 10,
Let's count to number 10.
Let's count the flowers up to 10.
1, 2, 3, 4, 5, 6, 7, 8, 9, 10.

(Repeat chorus.)

20 One, two, buckle my shoe

One, two, buckle my shoe
Three, four, shut the door
Five, six, pick up sticks
Seven, eight, lay them straight
Nine, ten, a good, fat hen.

21 I stretch up tall

I stretch up tall on tippy toe,   *(Stretch upwards while standing on toes.)*
Then down to touch my feet I go.   *(Bend over and touch feet.)*
Up, up, up my arms I send,   *(Uncurl arms above head.)*
Then down to touch my feet again. *(Bend over and touch feet.)*

22 Draw a circle

(Sung to the tune of 'Pop goes the weasel'.)

In the air around we draw   *(Draw a circle in the air.)*
Like the wheels on a bus.
What have we drawn, do you know?
We've drawn a circle!

23 Draw a triangle

(Sung to the tune of 'Three blind mice'.)

Up the hill, to the top,	*(Draw a diagonal line in the air from bottom to top.)*
Down the hill, then you stop,	*(Continue drawing a diagonal line in the air from top to bottom.)*
Straight across to where it all began,	*(Continue drawing a horizontal line in the air.)*

A three-sided shape is what you've drawn.
Can you tell me what it's called?
A triangle!

24 Draw a square

(Sung to the tune of 'Twinkle, twinkle, little star'.)

From the bottom to the top,	*(Draw a vertical line in the air from bottom to top.)*
Straight across and then you stop,	*(Continue drawing a horizontal line in the air.)*
Now back to the bottom you go,	*(Continue drawing a vertical line in the air from top to bottom.)*
Across and stop; now you know,	*(Continue drawing a horizontal line in the air.)*

All the lines are just the same.
Shall we start and draw again?

25 Draw a rectangle

(Sung to the tune of 'Old MacDonald had a farm'.)

A rectangle has four sides,	*(Hold up 4 fingers.)*
With two long and two short!	*(Show both hands a long distance apart, then a short distance apart.)*
Yes, all four sides are not the same,	*(Hold up 4 fingers and shake head side to side.)*
There's two long and two short!	*(Show both hands a long distance apart, then a short distance apart.)*
There are two sides long, and two sides short,	*(Show both hands a long distance apart, then a short distance apart.)*
Here they're long, here they're short,	*(Show both hands a long distance apart, then a short distance apart.)*
Count them there are four, so ...	*(Hold up 4 fingers.)*
It's a rectangle, I know!	

26 Shape song

(Sung to the tune of BINGO)

There is a shape it's like a ring	*(Draw a circle in the air.)*
Round and round it goes, O!	*(Draw a circle in the air.)*
Round, round, round again!	*(Move both arms in the motion of wheels on a train)*
Round, round, round again!	*(Move both arms in the motion of wheels on a train)*
Round, round, round again!	*(Move both arms in the motion of wheels on a train)*
And circle is its name, O!	*(Draw a circle in the air.)*
There is a shape that has three sides	*(Hold up 3 fingers.)*
Up, down, back it goes, O!	*(Hold up 3 fingers.)*
Up the hill, down and back!	*(Draw a triangle in the air.)*
Up the hill, down and back!	*(Draw a triangle in the air.)*
Up, down, then go back	*(Draw a triangle in the air.)*
And triangle is its name, O!	*(Draw a triangle in the air.)*
There is a shape that has four sides,	*(Hold up 4 fingers.)*
And all the sides are the same!	*(Hold up 4 fingers.)*
Four sides are the same!	*(Draw a square in the air.)*
Four sides are the same!	*(Draw a square in the air.)*
All four sides they are the same!	*(Draw a square in the air.)*
And square is that shape's name, O!	*(Draw a square in the air.)*
There is a shape that has four sides,	*(Hold up 4 fingers.)*
Those sides are not the same though!	*(Hold up 4 fingers.)*
Two long, just like this	*(Show hands a long distance apart.)*
Two short, just like this	*(Show hands a short distance apart.)*
Two are short, two are long	*(Draw a rectangle in the air.)*
And rectangle is its name, O!	*(Draw a rectangle in the air.)*

27 Clap your hands

Clap your hands, clap your hands	
One, two, three.	
Clap them, clap them	
Just like me.	*(Clap a rhythm for children to copy.)*

28 Seven days in the week*

(Sung to the tune of 'Oh my darling, Clementine!')

Always seven,
Always seven,
Seven days are in the week.
Every week there's always seven.
Seven days are in the week.

Sunday, Monday,
Tuesday, Wednesday,
Thursday, Friday,
Saturday!

Sunday, Monday,
Tuesday, Wednesday,
Thursday, Friday,
Saturday!

29 Sing with me the days of the week*

(Sung to the tune of 'Frère Jacques'.)

Every week
Has seven days,
Sing with me,
Sing with me.
Sunday, Monday, Tuesday,
Wednesday, Thursday, Friday,
Saturday.
What's today?

* You may wish to alter these rhymes to reflect the start and finish days of your week.

Generic games

Pairs and *Concentration* are almost identical games. The difference is that in *Pairs* children begin with the cards face **up** on the table in front of them. Whereas in *Concentration* the game begins with the cards face **down** on the table.

It is recommended that children begin by playing *Pairs*. When they are confident with the mathematics being practised, they can move on to *Concentration*.

Pairs

Set-up

- Shuffle all the cards. Spread them out face **up** on the table.

Number of players

2, 3 or 4

How to play

- Children take turns to look for two matching cards. They pick them up and show them to the other children.

- If the other child/children agree that the two cards match, the child puts the two cards beside them.

- If the cards do not match, they put the cards back in the same place that they found them.

- The next child then has a turn at trying to find matching cards.

- The game continues until all the cards have been matched.

- Children then count how many cards they have. The winner is the child who has the most cards.

Concentration

(Also known as *Memory* or *Pelmanism*.)

Set-up

- Shuffle all the cards. Spread them out face **down** on the table.

How to play

- Children take turns to choose two cards and turn them face up.
- If they match, the child puts the two cards beside them and has another turn.
- If the cards do not match, they put the cards back, face down, in the same place that they found them.
- The next child then has a turn at trying to find matching cards.
- The game continues until all the cards have been matched.
- Children then count how many cards they have. The winner is the child who has the most cards.

Notes

Notes

Notes